Robert. D. Elliot.

7 - July - 1990

Edinburgh

METAL FATIGUE

By the same author

Cement & Concrete Engineering
Cutting Tool Materials
A Dictionary of Alloys
A Dictionary of Heat Treatment (Publication arranged)
A Dictionary of Machining (Publication, 1972)
A Dictionary of Non-ferrous Metals
The Grinding of Steel
Guide to Uncommon Metals
Iron – Simply Explained
Machine Tools
Marketing the Technical Product
Mechanics for the Home Student
Metal Wear (Publication, 1972)
Outline of Metallurgy
Saws and Sawing Machinery
Steel Castings
Steel Files and their Uses
Successful Buying
The Surface Treatment of Steel
The Testing of Metals
Welding

In Collaboration

Foundry Practice
The Heat Treatment of Steel
The Machining of Steel
Non-ferrous Metals
Stainless and Heat Resisting Steels
Steel Manufacture – Simply Explained
Welding Technology

METAL FATIGUE

Eric N. Simons

DAVID & CHARLES : NEWTON ABBOT

ISBN 7153 5526 0

Dedicated to Albert Laybourn

COPYRIGHT NOTICE

© ERIC N. SIMONS 1972

All rights reserved. No part of this publication may be reproduced, stored in a retrieval system, or transmitted, in any form or by any means, electronic, mechanical, photocopying, recording or otherwise, without the prior permission of David & Charles (Publishers) Ltd

Set in 11 on 13-point Baskerville
and printed in Great Britain
by W. J. Holman Limited Dawlish
for David & Charles (Publishers) Limited
South Devon House Newton Abbot Devon

CONTENTS

                                page

List of Illustrations
Preface

Chapter 1   The Mechanism of Fatigue   11

       2   Metallurgical Factors - Effect of High   25
Temperature - Microstructure - Causes of
Failure - Corrosion Fatigue - Environmental
Effects

       3   Metals for Fatigue Resistance   59

       4   Behaviour of Particular Components   79

       5   Testing for Fatigue - High Temperature   92
Tests

       6   Fatigue Failure   108
Modern USA and USSR Research into

Glossary   Basic Fatigue Terms   116
              Metallurgical and Other Terms   120

Bibliography   124

Index   126

LIST OF ILLUSTRATIONS

Plates
	page
Fracture Surfaces Showing Unusual Patterns *(Ministry of Defence)*	17
Lug Failure *(Ministry of Defence)*	18
Fracture of Bolted Joints *(Ministry of Defence)*	35
How to Interpret Fracture Photographs *(Ministry of Defence)*	36
Fractured Face of Shaft *(Sheffield Testing Works Limited)*	53
4340 High Tensile Steel Component *(Ministry of Defence)*	53
Amsler Electrohydraulic Machine *(Ministry of Defence)*	54
Schenk Machine *(Ministry of Defence)*	71
Dowty Electrohydraulic Machine *(Ministry of Defence)*	71
Fracture Surface of a High Strength Steel *(National Physical Laboratory)*	72
Electronfractograph of Fatigue Striations in a High Tensile Steel *(National Physical Laboratory)*	72

In the Text
1	Fatigue Crack Growth	15
2	Severity Ratings for Inclusion Contents	19
3	Notch Effects on Tensile Strength	21
4	Cubic Space Lattice	31
5	Effect of Impurities	32
6	Effect of Nitriding on Fatigue Strength	47
7	Typical S-N Curve	55
8	Diagram of Stress Range for Cast Iron	70
9	Self-aligning Interference-fit Bolts	84
10	Stress Pin Fastener	86
11	Effect of Temperature on Superalloy Fatigue Strength	90
12	Fatigue Range and Creep Tests of Bolt Connections	90
13	Scatter in Fatigue Life Tests	94
14	Ultrasonic Tests and Cleanliness	100
15	Effect of Polarized Light	109

PREFACE

It is essential to fully understand fatigue, for relatively few constructional parts are immune to this phenomenon. The characteristic of modern engineering is an increase in operating stresses, temperatures and speeds. In addition, excess metal must be eliminated to reduce cost; weight must also be reduced to a minimum with no impairment of physical properties. These requirements are especially important in the design of aircraft, rockets, space missiles, etc. These developments have made the fatigue properties of metals more significant than their ordinary static strength properties, except where high temperatures are involved.

It is said that fatigue causes at least eighty per cent of the failures in modern engineering components, and the suddenness of these failures, both in the past and present, means that serious losses of life result from the absence of warning.

Despite all research work, some of which the following pages summarise, a complete basic explanation of fatigue has not yet been achieved, though it is fairly well established that local plastic deformation of a metal is less likely when the metal is in a fully elastic state. Fatigue is basically a gradual deterioration of the grains or crystals of a metal repeatedly stressed.

The present book does not investigate the numerous theories put forward, often only to be disproved, for example that of attrition, but indicates briefly what fatigue is and the factors affecting failure from this cause, the influence of techniques on fatigue resistance, the behaviour of metals under fatigue stresses and of different components under loading, the techniques of fatigue testing, and some results of modern research. A bibliography enables the reader to pursue the subject in more massive works, and the glossary gives the special terms used in connection with fatigue and its metallurgy.

I thank the American Society for Metals for valuable literature and the Sheffield Science and Commerce Library for books. Particularly I appreciate the help of Miss Angela Allott, Librarian of this section, for her co-operation. 'Metal Progress' is also thanked for some of the illustrations.

I believe no short, simple book on fatigue has previously been produced, and I hope that it will, until succeeded by a better, be of use to engineers, students, technicians, foremen, metallurgists, research workers, technical colleges and workers in the aircraft industry. In the examples, more weight has been given to this industry than to others owing to the many lives dependent on the ability of metals to withstand fatigue stresses over long periods of time.

Eric N. Simons
Eastbourne 1971

CHAPTER 1
THE MECHANISM OF FATIGUE

Metal fatigue is a comprehensive expression designed to indicate how metals behave when subjected to frequently repeated stresses over a particular range. More closely it is definable as the progressive deterioration of a metal until at a certain point it breaks, either completely or partly, by reason of alternating or other stress. The stress range involves repeated cycles of direct, bending, torsional or other stresses combined. The maximum stress of the cycle is numerically less than the particular stress leading to fracture after a single application. The ability of a metal to withstand stress is largely governed by the manner of its application. For example, if a metal wire is bent to and fro a sufficient number of times it will eventually break. Consideration of fatigue calls for the understanding of a number of special technical terms and metallurgical definitions, which will be found at the end of this book (page 116).

The higher the tensile strength and hardness of a steel or other metal, the greater is its fatigue resistance in most instances. Temperature also has an effect on fatigue strength and fatigue limit, especially where martensitic steels are concerned. Surface condition is another factor affecting fatigue resistance, and in general metal having a highly polished surface will be less affected by fatigue than one left as it was after being cast or hot worked, as by rolling or forging. Heat treatment and cold working also influence fatigue resistance and hardness. Assuming no great difference in composition exists, one metal at room temperature will not be greatly different in fatigue properties from another as long as their hardnesses are similar and unchanged. Some microstructural constituents, such as ferrite, do, however, considerably lessen fatigue resistance.

Shock or superficial flaws, discontinuities of the metal such as blowholes in a casting, local concentrations of stress such as notches and other causes of discontinuity, all make fracture more likely and rapid.

Fatigue fractures start from the surface as submicroscopic or microscopic cracks, rapidly developing under fatigue stresses and penetrating into the cross-section until complete failure results. These stresses at their highest point have a value less than that of the tensile strength of the metal.

Fatigue cracking is the result of two distinct processes: (i) the initiation of the crack by the stress; (ii) the extension of the crack in length to a point at which its dimension becomes critical. It is believed today that it may be possible to stabilize the microstructure of the metal and ascertain the specific group of crystals or grains that show the minimum life under stress of fatigue type. If such groups can be discovered and eliminated, it will be possible to attain with great regularity the most trustworthy fatigue resistance in hardened cast steel parts, for example. Research into elements that may achieve this result is already proceeding.

Work hardening is specially important in metal fatigue. It is the increase in hardness occurring when a metal is subject in the cold state to deformation by mechanical working.

THE MECHANISM OF FATIGUE

Take two strips of thin spring steel, one soft, the other correctly hardened and tempered. When bent through the same angle, the soft strip will remain bent, that is, permanently or plastically deformed. The other will spring back when the bending stress is removed. During plastic deformation the grains of the metal are distorted, not by being forced out of place, but by a process of slipping along the crystal planes of the grains, like a pile of books sliding over each other after a sideways push.

During the slipping process the atoms in the crystal planes are removed from their usual positions. In a plastically deformed metal, the energy expended in achieving this is stored, so that a piece of plastically deformed metal in the cold state is of greater hardness than that of the soft, unworked metal. The potential energy and hardness of a cold-deformed metal are proportional to the degree of deformation or cold working, so that the greater the cold deformation, the harder the metal. Eventually no further slip can be achieved, and if further stress is applied, the metal will fracture.

Striations

When a component or sheet of metal fractures, it is essential to examine the fracture itself to ascertain if possible the cause of failure. A fatigue fracture is readily perceived, since it shows marks like those left on the sand surface by a receding tide. When examined by the eye alone it will be seen that the area in which they occur is comparatively smooth and flat, with no sign of either contraction or elongation. It is therefore necessary to look for this smooth surface, which normally extends over a considerable area or areas of the original cross section. When it has been located, it is usual to find a second area over which final fracture is found when a particular load has been applied once only. The ribbed marks start at a point indicating the beginning of the microfracture and they correspond to the successive arrest of each crack with variation of stress. Parallel longitudinal smooth lines or ridges, concentric about the starting point of the crack, are termed *striations*, and begin after a few thousand stress cycles. Their appearance becomes clearer and clearer, and also more definite, as additional slip bands occur. The steps or terraces produced on the polished surfaces of metal crystals by their slipping over one another during stress are formed and simultaneously become broader. As the stress cycles continue, the striations broaden increasingly until the slip bands take up the whole surface or a crack begins before the load has become notably severe. The crack crosses the crystal or grain boundaries, but does not run along them. That portion of the cross section that fails is often coarsely crystalline or coarse-grained, as the metallurgist terms it. (To explain why would take us into the remoter depths of metallurgy.)

Cyclical or repeated loading causes a small degree of plastic deformation of the metal. This is essential to the striation of the crack by slip bands, and usually takes place in the region of maximum stress, causing a microcrack, which gradually develops into a large crack. Fatigue stress produces the surface furrows or ridges, whereas steady stress produces the surface steps. The slip bands developed by fatigue stress are characteristically undulating and lacking in regularity, whereas those developed by steady stress are straight and parallel. The curved slip bands are broad, short and seldom free from interruption.

Torsional Fatigue Cracks

In torsional fatigue the cracks run at an angle of 90° to the plane of maximum tensile stress. This type of fatigue is often indicated by slip band pitting and is sometimes termed surface fatigue. The effect is of local pits or flakes, and depends on the number of cycles of stress to which a given unit of surface volume is subjected. These pits or flakes are often found in roller bearings or gearing that

THE MECHANISM OF FATIGUE

are required to undergo combined rolling and sliding motion or rolling alone. When high stress and low cycle fatigue are experienced and when loading is completely reversed rather than carried out in a single direction, the causes of fatigue may be numerous and increasing. In shafts or bars subjected to torsional stress the cracks usually run at an angle of 45°, but should a longitudinal seam or the type of

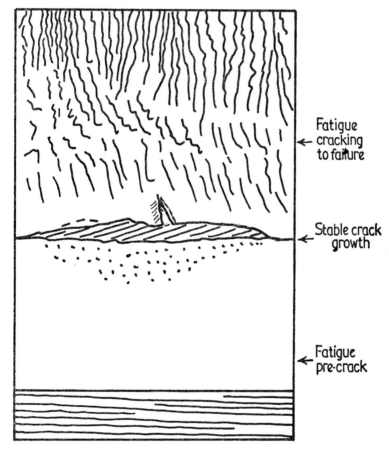

FIG 1 Fatigue crack growth. A crack under increasing load shows a bowed appearance

fibre structure produced by forging operations run longitudinally the crack may grow in the same direction.

When the cracks are not continuous in their growth under cyclic loading, the fracture resembles the concentric rings of the sawn timbers from a large tree. This appearance is, as earlier indicated, caused by the immediate halt of the progress of the crack at the termination of each cycle of stress. The rate at which the crack grows and its direction are governed by the grain boundaries; the presence or otherwise of nonmetallic or other inclusions; precipitates; the orientation of the crystals or grains; the levels of stress; etc. (See Fig 1). Plane bending usually gives rise to grooves or furrows.

Much valuable information can be obtained from the study of fatigue cracks, and especially the spacing and distribution of the surface rings or cracks. The closer together the striations, the more moderate the period of stress. It is also possible to determine whether the cycle of stress has been high or low. A state of fatigue giving widely-spaced striations, and broad rings frequently interrupted, is in most instances a sign of low cycle stress, and if a component showing this type of fracture has failed in service, it usually means that the number of cycles has been something below 5,000. If the fatigue fracture shows clearly defined and more crowded striations, the stress has been higher.

Corrosion Fatigue

Certain points must, however, be carefully noted. Corrosion fatigue (see later) does not easily show up, the striations being either hidden by the corrosion products, such as oxides, found when metal has fractured at high temperatures, or difficult to detect. In any event they are not in this instance an unchallengeable indication of fatigue. Again, some of the less hard metals, such as aluminium, and also the high strength steels of considerable hardness, may give fractures showing no striations at all. Cleavage in these by

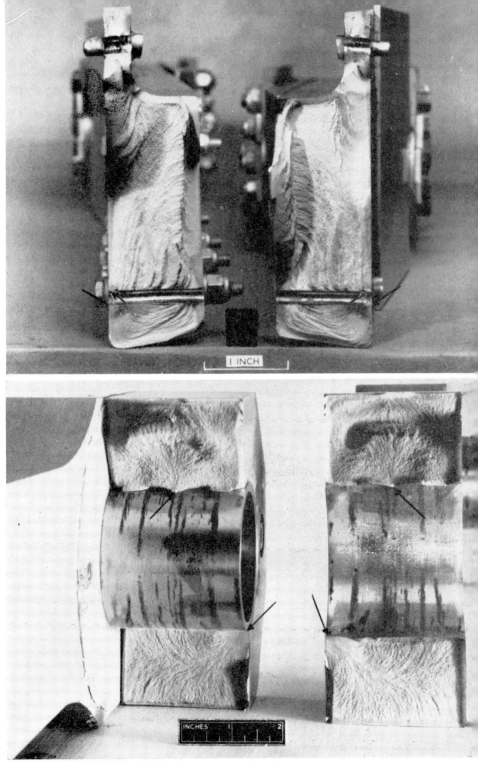

Page 17 Fracture surfaces showing unusual patterns

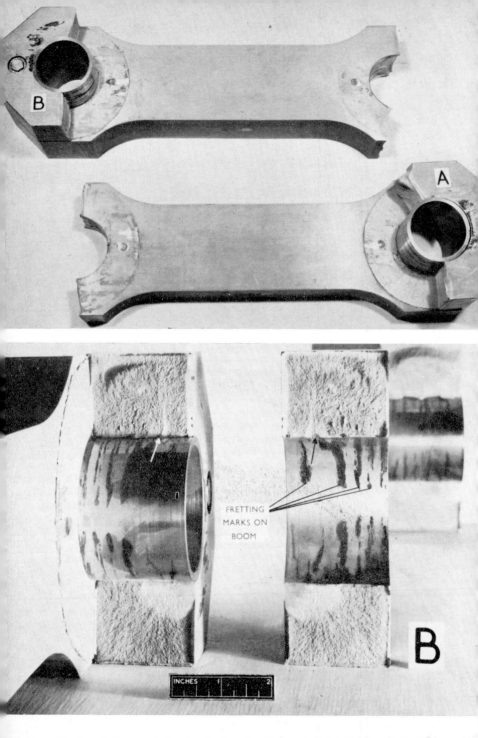

Page 18 Lug failure originating in fretting between interference fit bush and boom

fracture often runs along clearly defined crystal planes within the grains.

Nonmetallic Inclusions

Normally fracture begins and proceeds in this manner, but it may also occur through the cleavage of a nonmetallic inclusion, a hard precipitate, or by the opening of holes surrounding particles not firmly united to the metal. It follows, therefore, that the size, dispersion and type of particles composing the metal have a noteworthy influence on the ability to withstand fatigue fracture, and greatly influence the level at which energy is absorbed in the whole of the plastic region. Porosity previously present in the metal, and inadequate fusion of welds are other factors affecting fatigue fracture resistance. (See Fig 2.)

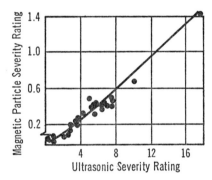

FIG 2 Severity ratings for inclusion contents: a linear relation exists for inclusion contents between the magnetic particle and the ultrasonic severity ratings (determined for the centre zones of an AISI 8620 billet)

Types of Fracture

There is a marked difference between fatigue fracture and fracture produced by static shear failure, at least as regards their appearance. Metallurgists use special names for different types of fracture. Thus, a bright and glittering fracture, in which the fracture develops along the cleavage planes of the separate grains, is termed a *crystalline* fracture, though the name is misleading to some extent, since all metal is of crystalline character, and no alteration in this type of microstructure is caused by repeated stress. Fracture takes place as the result of static shear after the cross section has lost strength owing to the extension of microscopic cracks between and within the metal grains.

Brittle metals such as cast iron, fractured with no deformation, show crystalline fractures somewhat resembling fatigue fractures. Where the metal has had its separate grains lengthened, the fracture is grey and dull, and characterises a ductile but not homogeneous metal such as wrought iron. This is termed a *woody* or *fibrous* fracture. Low carbon steel when fractured may show an extremely smooth, fine, dull grain, and this is termed a *silky* fracture. When the cleavage appears to follow the boundaries of the individual crystals, this is known as a *granular* fracture. Fine-grained material of low ductility gives in many instances a type of fracture termed *vitreous* or glassy. Where a fracture appears in a material of ductile character greatly deformed, it is known as *tough*. *Brittle* fractures are those in which the metal fractures as a result of cleavage owing to the cohesion having been exceeded.

Fatigue cracks, whose appearance has been described, may be caused by non-metallic inclusions or other flaws, but mostly they are the result of notches, which act as stress raisers. Because of this, any component or surface of metallic character likely to be subjected to a large number of cycles of rotating or reciprocating motion, or loading of dynamic type, must be safeguarded against the development of stress concentrations leading to these superficial flaws. Heavy machining incisions, gouged areas, indentations, accidental notches, cannot be entirely eliminated, but may be minimised by choosing the right metals for particular conditions, especially those intended to keep nonmetallic inclusions out of the microstructure or to prevent faulty surface finish. Highly polished work is less likely to show the marks of cutting or other tools, and will also be more carefully treated by the operators. Under the general heading of 'notches' are included sharp corners, grooves, fillets, radii, diametral holes in round bars, plates, etc. These are what are termed above 'stress raisers'. (See Fig 3.)

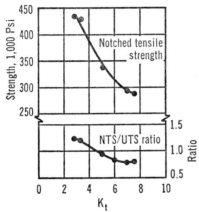

FIG 3 Notch effects on tensile strength: the through tensile strength drops as notches become sharper; the ratio remains above 1 until the K_t reaches nearly 5. (Samples were solution treated for 1hr at 1473°F, air cooled and aged 3hr at 950°F)

A common source of fatigue cracks is an oilhole or screwhole or some abrupt change of cross section at a particular point. Local corrosion may set up surface pitting, which is also a focus of crack formation, so that it constitutes yet another argument in favour of high polish. Freedom from pitting of this type greatly enhances fatigue resistance, and should corrosion coincide with fatigue, the fatigue limit may decline by up to 65%, with highly injurious results, the corrosion and the fatigue being speeded up.

In parts whose function involves motion, or which are subjected to dynamic stresses, it is common for some portion of the cycle to exist in which the faulty zone is the only one to carry a considerable part of the load. When this happens, the zone is likely to show brittleness, perhaps because of the swift deformation of minute portions, so that a crack forms and gradually propagates until fracture ensues. Once the crack has begun, the rest of the sound uncracked metal has to accept a proportionately greater degree of stress, so that the development of the cracks becomes more rather than less rapid, as contrasted with static loading leading to work hardening. The existence of these flaws consequently lessens the effective fatigue limit of the material.

Broadly a metal of soft character is not so greatly affected by notches as a harder metal, irrespective of the origin of this softness. Because this lower sensitivity is likely to be caused by the more easy manipulation of the softer material, the precise ability to work-harden, as compared to that of the harder material, has to be taken into account. Consequently while virtually all the soft metals are, as stated, less sensitive, with steels this is not invariably true, inasmuch as a hard and relatively brittle steel could be softened by heat treatment so as to show considerable ductility. The required properties appear to be a reasonably low yield strength and ability to be worked, enabling a good deal of plastic deformation to spread the concentrated stresses as widely as possible over the affected area, so preventing local areas from becoming brittle and cracking.

Internal stresses, of which nonmetallic inclusions and holes are examples, are to be avoided wherever possible, partly because of the natural concentrations of stress caused by discontinuity of the solid metal, and partly because the inclusion may be less able to 'give' than the metal. Thus, the stress is even more concentrated than by a hole alone. For this reason it is becoming the practice, particularly in

THE MECHANISM OF FATIGUE 23

the United States of America, to indicate a specific inclusion rate so as to govern this factor.

Treatment of Fatigue Cracks

Fatigue cracks may sometimes be eliminated at an early stage by machining away the surface layer. The cracks are readily initiated by dislocation of metallic atoms and lead to great depth of the slip band groove. Disturbance of the surface layer, produced by the mechanical impingement of metal shot (shot peening), rolling, or other intentional working treatment, gives the metal a superior resistance to fatigue.

Preventing Fatigue Failures

The principal requirements for the prevention of fatigue failure have been summarised in earlier passages. First, the amount of stress to which a member or component is subjected must not be greater than the fatigue strength. The size of those cross sections regarded as critical should be increased, and the dimensions of the particular parts are best determined by exchanging needless material performing no valuable purpose in noncritical components, for metal giving higher strength in those subjected to higher stress.

Maximum trustworthiness of any construction or machine is achieved when every part is identical in safety factor irrespective of its mode of failure. In practice, the safety factor for each individual part and mode of failure should be slightly modified, upwards or downwards, from the nominal overall design as governed by comparative cost and the significance of possible failure. In other words, the safety factor must be somewhat higher than for other modes of failure, because failure of this type occurs with no advance indication and frequently results in great injury to the system as a whole.

All mean stresses are to be regarded as 0 (zero). If any mean stresses are not zero, the corresponding values of nominal stress have to be worked out from the fatigue operating strength diagram. Most fatigue failures are caused by inadequate fatigue properties. The differences in their responce to notches or to other surface blemishes have to be borne in mind. In particular it should be noted that cast metals are less variable in this respect than wrought. Practical design points to be taken into account are that notch radii should be increased, and fillets undercut, so that the flow lines may be less sharp. This is not to be taken as relating to those superimposed notches harmful to the fatigue strength of the metal. Stresses should if possible be redistributed and mean stress reduced, while a modification of the form of central cross sections may sometimes be advantageous.

CHAPTER 2

METALLURGICAL FACTORS - EFFECT OF HIGH TEMPERATURE - MICRO-STRUCTURE - CAUSES OF FAILURE - CORROSION FATIGUE - ENVIRONMENTAL EFFECTS

METALLURGICAL FACTORS

Much progress has been made in the study of metals operating under alternating loads, but despite a considerable quantity of experimental evidence in regard to ductile metals, the fatigue of hardened steels has been only partly studied. The problem is, however, of importance for numerous kinds of steels possessiong high strength, and if their behaviour and trustworthiness in cyclic stressing could be established, great advantage would result. The tests on hardened steels reveal that they differ in behaviour from ductile metals, and it is evident that the way in which fatigue failure is brought about is somewhat different from that seen in ductile constructional materials. In ductile materials tested under fatigue conditions, three principal stages of damage are apparent: (i) the primary stage up to the initiation of the first crack; (ii) the extension of the cracks, that is the period during which the first fatigue fracture area is developed; and (iii) the moment of final fracture. For hardened steels, the fatigue curves show that a continuous relation within the range 500-1000 million cycles is visible between the fatigue strength and the number of cycles of load application over a period of time. This relation is expressible as a straight line in s—log N coordinates. Variation of flexure during cyclic stressing shows that hardened steel fails through fatigue somewhat differently from ductile materials.

Fatigue failure usually occurs microscopically, either with small strains and cracking, or by shear after a great amount of plastic strain. Brittle static fracture is normally the result of large stresses or maximum elongations. Fatigue failure usually takes place in a brittle manner, but beyond question it is associated with shear stresses. Failure resulting from cyclic contact stresses is characteristic of balls, rollers and cages of roller bearings, gears, rails, wheels, steel tyres, journals of cold rolling mills and many other parts. Cyclic loads giving rise to variable stresses cause intersection of dislocations, producing many vacancies. The vacancies build up into clusters and form submicroscopic pores and cracks. This mainly causes a rise in the hardness of surface layers during contact loadings.

Some investigators consider the statement that the notch sensitivity of a steel always increases with strength under cyclic loads is inaccurate. They argue that the strength and hardness of the steels are not the main causes for notch sensitivity, and that there is no relation between the two phenomena. The reason for the low notch sensitivity of a high strength steel is believed to be its increased damping, which arises from the heterogeneous microstructure. This increases with damping and reduces the sensitivity of the steel to concentrations of stress. Experiment, they claim, has indicated that workhardening produces no change in the general relation between notch sensitivity and strength, but affects only the absolute value of the sensitivity index. Workhardening in the surface layer of the notch, caused by machining, is not the explanation of the reduced sensitivity of high strength steel.

Damping Capacity

Damping is a term used in connection with vibrating mechanical systems to indicate the way in which certain influences limit and reduce the degree of vibration by resisting the movement and taking up the energy from the system. A vibration that was not damped would continue indefinitely, but as this state of affairs is non-existent in nature, all free vibrations must eventually cease. Alternative names for damping are crackless plasticity, elastic hysteresis, internal friction, and mechanical hysteresis loss, but some of these involve a certain amplification of the term.

Damping capacity is the extent to which a metal is able to absorb or reduce vibrations during a fully reversed stress cycle. This ability seems to affect fatigue resistance, but there is considerable controversy regarding the relation between these two properties. Fatigue failure mainly occurs because at a constantly repeated stress above the endurance limit, the damping increases in effect and rate until the metal fractures. In other words, the metal takes up internally ever more vibrations until the stresses eventually become great enough collectively to initiate a submicroscopic crack. When, however, the cyclic stress is less than the endurance limit, damping reaches a constant value, that is, work hardening is great enough to spread out the concentration of stress, which is consequently halted or declines, owing to the stress being lower than the elastic limit. Thus, no surface hardening by work occurs, the stresses being elastically taken up.

Certain metals of high damping capacity, including cast iron, can resist notch effects, but are likely to be affected by overstress, so that their endurance limit will be low and they are extremely unlikely to have this limit affected by surface flaws and notches. Those metals having low damping capacity, however, embodying most steels subjected to heat treatment, will withstand overstress, but are likely to be highly sensitive to surface blemishes. Low damping capacity metals should, therefore, be chosen for components needing high resistance to alternating stress cycles, and must be given a high degree of surface finish. Materials of high damping capacity, on the other hand, do not need so fine a surface finish.

Some machine components such as high speed lathe beds or engine crankshafts, have to withstand vibrational loading. It might be thought that these would demand a material of high tensile strength alone, if the design of the machine were the sole guiding factor; but where the material has a low damping capacity, the vibrations may aggregate to a considerable magnitude, so that fracture would take place much more rapidly than if the metal had a lower tensile strength but a higher damping capacity.

EFFECT OF HIGH TEMPERATURE

Thus far we have considered only metals at room temperature, but there has developed of recent years a great deal of interest in fatigue at high temperatures, simply because technological development has led to the introduction of an increasing number of engineering structures in which these temperatures are encountered, such as gas turbines, diesel engines, nuclear reactors, etc. Fatigue is experienced in such machines in various forms. For example, rotation may induce vibration, and so may fluid motion, so that fracture occurs by reason of a large number of stress cycles. Fatigue also arises because of the start and stop sequences, and is usually termed thermal or low cycle fatigue. Strain is the regulating and varying factor, and results from temperature influences.

Typical difficulties arising from high temperature include gases or liquids which by producing surface reactions promote the formation of cracks, increase their growth, and eventually cause fracture. Long holding time between cycles causes creep influences to combine with fatigue so that whereas in the normal process the crack is no longer propa-

METALLURGICAL FACTORS

gated from the more ductile across-the-grain process, the crack is propagated by the less ductile between-the-grain process.

In addition there is always the possibility that the metal's properties may vary when the temperature is maintained for a considerable length of time, largely because of the phenomenon known as ageing, a change in the properties of a substance with time, usually an increase in tensile strength and hardness. Such a change may lead to instability of phase. (Phase, in the language of metallurgy, signifies that portion of a system which is chemically and physically homogeneous.) Another possible harmful result is creep.

Finally, cycling at high temperature is likely to make the forecasting of stresses and strains more intricate and difficult, while the interaction of temperature cycling and strain cycling may also lead to a lack of certainty.

All these complexities are a foundation whereby it is feasible to recognize those regions in which fatigue poses questions needing answers. These are the constant subject of research by metallurgists, mechanical engineers, and chemists, especially those problems involving the influence of the temperature maintenance period, in which the strain is sustained at one and the same level for specified periods of time during the cyclic strain.

As will be seen later, the methods adopted in testing are of great importance when investigating the effects of fatigue in metals at elevated temperatures.

The Cyclic Stresses

Fatigue fractures may be caused by repeated cycles or repetitions of single or combined stresses, and become more rapid when conditions of shock or surface notches are present, as well as other conditions earlier specified. There are three primary kinds of cyclical stress: (a) alternating, (b) pulsating, and (c) fluctuating. Whatever the stress cycle, it is expressible as $M \pm \frac{1}{2}R$, where M signifies the average stress and R the stress range. Stress range is the maximum range applicable for an indefinitely great number of repetitions not producing failure.

The number of cycles required to produce fracture depends on the stress applied, so that the heavier the stress, the smaller the number of cycles required to cause failure. As stress declines, the number of cycles necessary to produce fracture increases, until in the end a stress is attained that will not produce fracture within the usual test limits. The period of time for a standard test of this type is governed by the metal undergoing investigation and the kind of loading applied. For the best results it should be a minimum of about 500,000 cycles for extremely hard steel, 5,000-10,000 for mild steel, 10^6 for cast steel and cast iron, and from 10^6 to about 5×10^7 for the nonferrous alloys according to their type. It appears to be the case that if no fracture is developed inside these limits, it will not be produced at all.

The failure of metals by fatigue, under rapidly alternating stresses, is largely a matter of stresses within the individual grains. The study of single crystals under strain and fatigue has led to many theories as to the cause of fatigue failure, but it appears to be established that the fatigue range is related to the ultimate strength of the metal, and is not connected with the elastic limit of the material. Fatigue failure is basically a gradual deterioration of the grain microstructure of a metal, as earlier stated.

The fatigue life of a component expressed in stress cycles is notably influenced by contact stress, and in roller bearings, for example, it is inversely proportional to the 9th or 10th power of contact stress. Reduction of stress is possibly the most important factor in the life of a component under fatigue conditions. For example, the existence of a film of lubricant on a bearing surface changes the pattern of stress operating on a specified unit volume and, therefore, alters its fatigue life.

MICROSTRUCTURE

Metallurgical Examination

After the fracture has been studied by the eye, the next step is determination of the more easily recognisable, variable factors of metallurgical type, such as the metal surface, grain size, the nonmetallic particles present as inclusions, their form and their orientation to the maximum tensile stress, and finally any perceptible effects of the environment. Grain size and nonmetallic inclusions are dealt with later. Another factor of considerable significance is high stress, low cycle fatigue. Often it is the level of cyclic strain rather than stress that decides the likelihood of future failure from fatigue.

To render the pages that follow more intelligible, let us consider cleavage planes and crystal boundaries. The atoms of a metal are located in several sets of parallel planes. In the crystals or grains of which metals are composed, the regular cubic arrangement of these atoms produces planes either parallel with the three axes of the cube or diametrical, and along these planes atoms can more easily slide over one another.

These are the cleavage planes. At the junctions or boundaries of the different grains there are, of necessity, spare

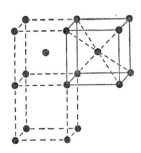 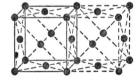

Fig 4 Cubic space lattice: the 'face-centred' cubic lattice of γ iron *(left)* and the 'body-centred' cubic lattice of α and β iron *(right)*

atoms not occupying regular or symmetrical positions on the space lattice, that is the dimensional geometric pattern in which the metal atoms arrange themselves, and upon which the crystals are built. There are not enough atoms to form the complete cubic arrangements shown in Fig 4. The surplus atoms will obviously occupy random or indeterminate positions, and this haphazard arrangement actually forms the crystal boundaries, that is, the atom arrangement in the boundaries is irregular.

In the boundaries there are no cleavage planes, so that they do not split so readily as these grains themselves and are, in fact, stronger. That is why fracture most easily occurs along cleavage planes. Moreover, the smaller the grains, the more boundaries have to be broken across for fracture to occur. A fine-grained microstructure is therefore always stronger than a coarse-grained structure of the same metal or alloy.

When impurities or other constituents are present, the story is entirely different. During cooling, a metal throws out the impurities or other foreign atoms. These may then form layers or films separating the grains of the parent metal from each other. Fig 5 (b) shows roughly how this

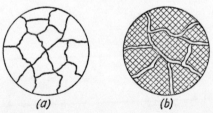

FIG 5 Effects of impurities: Crystal thrown layers

appears. The cross hatched areas indicate the pure metal or solid solution, the white streaks represent the rejected substance or impurity.

The important conclusion can therefore be drawn that the strength of a metal or alloy is largely the strength or fragility of its grain boundaries. With a pure metal of uniform solid solution the boundaries are stronger than the grains themselves, and fracture of the ordinary kind takes place across the crystals along the cleavage planes. When, however, the boundaries contain impurities or other brittle constituents, the fracture may occur along the grain boundaries or separating surfaces. Brittle fractures of this type are somewhat similar to those produced by fatigue.

Grain Size

Grain size must be understood. It is not possible here to delve into the highly complex and intricate research work devoted to grain size and its significance. Certain results of this work bear, however, upon the subject under discussion. It is known that smaller grain size, except in special circumstances indicated later, is always associated with greater toughness and strength in a metal. The final grain size in steel is affected by the temperature at which the steel stood when it began to cool. The higher the temperature when cooling begins, the larger and coarser the grain of the final product. Just as there is a relation between toughness and grain size, so there is a relation between grain size and the temperature from which cooling begins.

However irregular, coarse and large the grains may be, when a steel is heated to a temperature known as the upper critical point, a specific temperature and pressure are achieved at which changes of phase are soon apparent in steel. The large grains disappear and their place is taken by grains of the best possible size. It is now known that to heat a steel for a considerable period above 1000°C (1830°F) greatly enlarges the grains and embrittles the steel. Thus, prolonged overheating can do irreparable harm.

Cleavage Planes

The cleavage planes of iron and steel do not always lie in the same direction, but can lie at all angles to each other, this ensuring the strength for which they are valuable. Nevertheless, while fine grain usually means good quality, it can sometimes be harmful.

Grain size is often measured in Britain, and almost always in the United States, by the valuable McQuaid-Ehn carburizing test, obtained by testing the steel and expressing the result as a formula by a series of American Society for Testing Materials numbers. From Table I it can be seen that as grain size number increases, the size itself lessens, while for a particular hardness or tensile strength a tougher steel is indicated. Steels with grain size numbers (n) of four and below are termed coarse grained, and from five upwards as fine grained.

It must be emphasised, however, that all steels, irrespective of their grain size, are subject to grain growth if overheated to a considerable degree. Some steels, especially alloy steels, are inherently fine grained, but most carbon steels fall into coarse and fine categories according to the degree of oxidation of the melt by elements such as aluminium and vanadium. Such steels are termed of 'controlled' grain size.

TABLE 1

Value of n	Grains/in$^2 \times 100$	Grains/mm$^2 \times 10^{-1}$
2	2	$\simeq 3$
3	4	6
4	8	12
5	16	24
6	32	48
7	64	96
8	128	192

This table is based on n, the grain size index or number, so that when $n = 1$, there are $2^{1-1} = 2^0 = 1$ grains/in$^2 \times 100$ and therefore the average grain size is 0.01in.

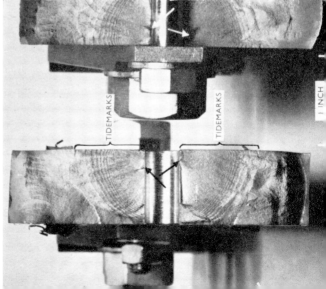

Page 35 Fracture of bolted joints showing 'tidemarks'

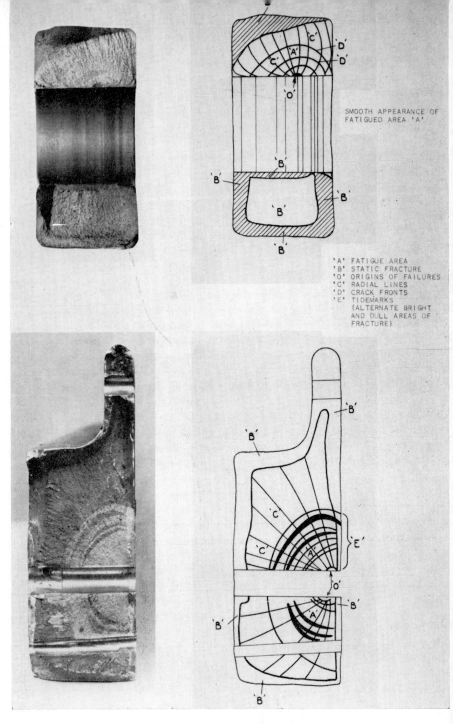

Page 36 How to interpret fracture photographs

CAUSES OF FAILURE

Notch Sensitivity

Internal defects of a metal, such as blowholes and non-metallic inclusions, have been seen to be primary factors in fatigue failure, not only because they introduce discontinuities into the metal, but also because inclusions are in themselves less plastic on the whole than the metal itself, and cause even greater stress concentration. It is therefore desirable in many instances to indicate that the metal should have a specified limit of inclusions. To detect inclusions of greater size, magnetic particle inspection is employed, and steels, in particular, are often chosen with this as a basis (see also page 39).

There was for a long period a belief that the notch sensitivity of a steel always increases as its strength under cyclic loads increases, but this is not so. The strength and hardness of the steels are not the main cause of notch sensitivity, and in fact no direct relation between the two phenomena exists. It has been shown by experiment that for some steels notch sensitivity at first increases strength appreciably, but the strength declines considerably at an ultimate strength of $1,250 MN/m^2$ (81.7 ton/in^2). The sensitivity of a high strength steel is even a little lower than that of low strength steel ($950 MN/m^2$).

Increased damping appears to be the reason for the low notch sensitivity of high strength steel and is said to be the result of a heterogeneous microstructure, which reduces the steel's sensitivity to stress concentrations. It seems that a closer study of the heterogeneous microstructure might throw more light on the cause of increased damping. According to experiments, workhardening does not change the general relation between notch sensitivity and strength, but affects only the absolute value of the sensitivity index. Workhardening in the surface layer of the notch caused by machining does not explain the reduced sensitivity of high strength steel.

Stress Concentrations

The need for accurate calculations in machine design when variable stresses have to be encountered has brought into focus the effect of stress raisers on fatigue strength. Studies have confirmed that for a range of steels investigated, strength is increased under conditions of varying load amplitude. Cumulative damage was found to be governed by the stress level, normally decreasing as stress increased. Such damage remains constant only in special cases in which certain definite conditions are fulfilled. A definitive relation between strengthening and the effective stress concentration factor has been observed. For heavy concentrations of stress and high overloads, strengthening may revert to weakening. Resistance to overloading is governed by the working conditions of the metal, which must be identical before true comparisons can be made. It is reasonable to suppose that effective stress concentration factors for both short and long service lives are virtually independent of the loading conditions.

Defects

Among the surface blemishes and defects of internal type that constitute major factors in fatigue failure are blowholes, the round or elongated smooth-walled gas-filled cavities in solid metals. They introduce discontinuities into the solid metal, and facilitate stress concentration. Moreover, nonmetallic inclusions, sometimes restricted in number, cause even greater concentrations of stress. Inclusions are produced by deoxidation reactions, that is, oxides, silicates and sulphides that have not been able to coalesce and rise to the surface of the molten steel. They may also be fluxed refractories and slag, introduced during stages of tapping and casting.

METALLURGICAL FACTORS

Inclusions are more potent in their effect on the fatigue life of carbon steels and alloy steels than chemical composition, microstructure, or gradients of stress. Their character, number, dimensions and dispersion all play a part, and to prevent them, the modern tendency is to melt metals by vacuum processes, but this is expensive, so that it can be used only in a small number of highly specialised fields, such as rocketry, aerospace projects, etc.

Magnetic Particle Inspection

Magnetic particle inspection has been used as a nondestructive means of determining the existence and extent of inclusions in ferromagnetic metals. This involves the use of finely divided magnetic particles applied to the magnetised metal either by spraying or brushing. These particles are attracted to and outline the pattern of magnetic leakage fields created by discontinuities. The pattern is usually shown on an oscilloscopic screen.

The basis of the method is that a magnetic field is built up in and around the component being inspected, and if a crack or discontinuity is present, north poles will be formed on one edge and south poles on the other, thus generating magnetic field lines. A magnetic fluid or detecting ink consisting of a light oil vehicle, such as paraffin, contains a proportion of small particles of magnetic iron oxide in suspension. The particles have a magnitude of only a few microns. Application of the fluid to magnetised metal causes magnetic poles to appear at discontinuities, and the particles are drawn to and gather at the poles, so showing where the discontinuities lie.

Ultraviolet light is also used for the same purpose when inspecting the interior of tubes, the powder pattern showing up as greenish-yellow lines against a dark background.

Decarburization

Another agency responsible for fatigue failure, especially

in springs, is decarburization or 'soft skin'. However, this is less injurious to hot-rolled than to cold-rolled or coiled springs because after hot forming, other superficial blemishes may set up as many fatigue cracks as the soft skin of decarburization produces.

The degree of decarbuization of steels is much lower than than formerly because heat treatment furnaces have so greatly improved with the introduction of atmosphere controlled designs, as well as by the general improvement in heat treatment practice. Soft skin alone is seldom responsible for failure in springs, as other defects are much more likely to be the cause. If, however, inspection of the spring reveals a decarburized ferritic ring on the circumference of the wire, discarding the material is justified. Soft skin is of greater depth 125-375µm (0.005-0.012in) when the steel stock is wound hot from a large diameter, but this can be prevented by special manufacturing techniques.

A not unmanageable degree of soft skin can be remedied in some instances, and fatigue resistance improved, by either shot peening or treatment in a salt bath so as to restore some of the surface carbon lost. Shallow hardening will also give the type of stress pattern preferred, but this depends on a suitably designed spring. Shot peening pre-stresses the compressed surface of the spring, and is most effective when the blemishes are of no great depth on the surface. The minimum wire diameter to which it is applicable is, however, 1.58mm ($\frac{1}{16}$in).

Shot blasting is less advantageous for temperatures approaching 260°C (500°F), and is useless for temperatures above 420°C (790°F). The shot is directed towards the interior of the coiled spring where the stress reaches its maximum microstructure.

In connection with decarburization, the microstructural condition of the steel is important, but two terms must be understood; *hypoeutectic* and *hypereutectoid*. The first represents a steel with lower carbon content; the hypereutectoid has higher carbon content, the mean being pearlite carbon content, pearlite being the iron-cementite microconstituent formed by the simultaneous production of iron and cementite. Hypereutectic steel contains less than 0.9% carbon. A hypereutectoid steel, on the other hand, has more than this percentage. The hypereutectic steel is more resistant to soft skin decarburization than the hypereutectoid, especially when the springs are for clock, watch, or power applications.

Microstructure

Microstructure is also important in relation to the fatigue life of steels, and one aspect of this is the martensite content. Martensite is the micro-constituent of steel formed when it is quenched from temperatures above the upper critical point. It is made up of needle-like intersecting shapes, readily recognisable, but may also appear on occasion as a seemingly formless ivory-tinted constituent.

The higher the martensite content of a steel, the greater the fatigue limit and vice versa. The fatigue limit of medium and low alloy carbon steels is reduced from 5-8% by quenching in air rather than oil, and tempering to produce identical tensile strength. Aircraft crankshafts have been rendered more fatigue-resistant by this treatment.

Sulphide Inclusions

It is known that among nonmetallic inclusions harmful to metals are sulphides, especially at fillets, the roots of splines, threaded joints, etc, but on the whole little work has been done on their effect on fatigue. It appears that notched surfaces are more harmfully affected than smooth, but for some notched carbon steels with 0.40-0.48% carbon and 0.24 to 0.33% sulphur, the fatigue strength at 10^6 cycles is from 3-30% higher than for similar steels with 0.43-0.50% carbon, and the same sulphur content.

CORROSION FATIGUE

Metals subjected to corrosion in combination with stress display a considerably higher fatigue strength if all corrosive influences are eliminated. This is shown by the improved strength values obtained when a component is heated in a controlled atmosphere furnace having a neutral atmosphere. The type of fatigue caused by corrosion is termed corrosion fatigue, and is the effect of applying repeated or fluctuating stresses in a corrosive environment. It gives the metal a shorter service life than if the repeated stresses were accompanied by no corrosion, or if the corrosion environment alone were concerned, without the stresses.

Much depends on the kind of material affected. For example, stainless steels show a much greater degree of this type of fatigue than other steels, and here the chromium content is important. The greater it is, the greater the corrosion fatigue resistance. The less the chromium and the higher the nickel content, the lower the corrosion fatigue resistance. On the other hand, previous heat treatment shows a considerable influence. Thus, a chromium steel having 0.38% carbon and 14.5% nickel showed a greater amount of damage as represented by the ratio $\frac{\text{corrosion fatigue}}{\text{endurance limit}}$ when annealed, as compared to the same steel quenched and tempered using a salt water quench. On the other hand, when both samples were tested in ordinary water, there was no appreciable difference in the ratio.

Similarly, an austenitic stainless steel quenched in ordinary water was much more damaged than when quenched in salt water.

Corrosion Fatigue of Copper and Copper Alloys

Şimilar combinations of cyclic stress and corrosive agencies, such as normally produce pits on the surface of the metal, also produce corrosion fatigue. This is not the same as ordinary fatigue, because it is seldom that a fractured

METALLURGICAL FACTORS

component of these metals fails to show a multiplicity of cracks, whereas in ordinary fatigue it usually shows only one. These cracks, as earlier indicated, normally run at 90° to the maximum tensile stress in the zone affected. The united effect of the two causes of fatigue, when both occur at the same time, is much more harmful than that of either cause alone. The great advantage of the copper alloys and copper is that they show high resistance to corrosion fatigue even when the service involves repeated stress and corrosion. They are thus particularly suitable for such components as springs, switches, diaphragms, syphon bellows, pipes and tubes in fuel lines for aircraft and automobiles, condenser tubing and heat exchanger tubes, as well as for fourdrinier wire in the paper industry.

The higher the initial fatigue limit and corrosion resistance under the conditions of use, indeed, of any suitable use where cycles of stress and corrosion are both involved, the better will be the service life. The most commonly used copper alloys for work involving corrosion and stress cycles in combination are beryllium copper, phosphor bronze, aluminium bronze and cupro-nickel.

Corrosion fatigue in a wrought copper alloy is typically transcrystalline, and the cracks normally start in a rough surface. Spontaneous failure caused by cracks sometimes results from the combined effect of corrosion and residual and applied stress, and here the cracking is intercrystalline. This, however, is more properly termed stress corrosion failure.

Cause of Corrosion Fatigue

In essence, this type of failure results from the decline in strength a metal shows after it has been subjected to cycles of repeated stress in an environment of corrosive type. When no corrosion exists, though this is rarely the case, the fatigue limit has a reasonably exact value. Moreover, the corrosion that aids in producing corrosion fatigue is usually very small. It is, therefore, necessary to eliminate corrosion as far as possible at the outset, since as soon as it has started, the effects of cyclic stressing may result in speedy failure even if additional attack is eliminated.

Corrosion fatigue cracks are extremely difficult to detect before they have developed to the extent that causes fracture. For this reason, attention must be paid to the application of means to protect the metals during the whole of their service life. This result can be achieved with zinc alloys by means of electrochemical methods.

Corrosion Fatigue and Zinc Alloys

The most effective practice for the protection of these alloys is the application of a zinc coating to provide complete cathodic protection. Such coatings may have slight surface blemishes, but these are of little significance, and their use eliminates corrosion even at the 'growing points' or nuclei at which the crack is initiated. Such points are usually in the submicroscopic fissures in which corrosion fatigue is born.

It may be imagined that coating of any type would serve the same purpose, but this is not so. The coating must be anodic, especially to steels, because otherwise the protection is not sacrificial and does not hinder the initiation of corrosion fatigue in its surface blemishes. This would not matter if the possibility of fracture caused by fatigue had not to be considered in addition.

When this type of fatigue is likely, however, a sacrificial coating becomes indispensable, as otherwise the discontinuities in the coating applied may themselves set up harmful pitting, and this is specially troublesome when the area of corrosion is small, so that the development of fatigue cracks is enhanced. Otherwise, the small amount of corrosion sometimes found when an ordinary non-sacrificial coating is used would not cause much difficulty.

METALLURGICAL FACTORS 45

Zinc and Corrosion Fatigue

Wires subjected to cyclic stresses in marine waters are likely to fail quickly in service owing to the cyclic stresses occasioned by vibration. By galvanizing the wires this fatigue failure is prevented.

By combining galvanization and zinc anodes, steel has been protected from corrosion fatigue in many applications, whether at sea or in storage on shore.

Cause of Corrosion

The importance of zinc coatings will be better understood if the mechanism of corrosion is described. It has been found that corrosion is largely generated by local variations in the solution in which a metal is immersed. These variations are of two kinds:

(i) One part of the water or solution is normally richer in oxygen than the other, and, if so, the richer part will render any metal cathodic to that part not in contact with it. That is, the metal surrounded by water richer in oxygen acts as a cathode, while the metal surrounded by water less rich in oxygen acts as an anode. The anodic metal is therefore much more rapidly corroded than the cathodic. This explains why corrosion at the bases of cracks and other surface blemishes is accelerated.

An electric current passing through a solution, with the metal plate constituting the entry point or anode, dissolves the metal and deposits it on the cathode. Actually, this deposition is somewhat slower than might be expected, because if the metal is one that does not readily dissolve and part with its ions, not enough may be present in contact with the cathode, and deposition will not take place to any great extent. In these conditions hydrogen ions are freed instead. These collect around the cathode surface and screen it from deposition. If the water is rich in oxygen, it (the oxygen) combines with the hydrogen ions to form water. This tears holes in the hydrogen screen, electric current passes through the gaps and deposition starts up again. Hence, the oxygen-rich parts of the solution become cathodic to the oxygen-poor parts, which become anodic and corrode the metal more rapidly. There are always variations in the oxygen content of the solution or 'electrolyte', which absorbs oxygen from the air, so that as atmospheric pressure fluctuates, and also its temperature, there is unlikely to be a perfect balance between solution and air.

(ii) If more metallic ions collect at one point than at another, this collection becomes a cathodic region, at which point deposition is probable. In consequence, wherever the metallic ions are thinly clustered in contrast to other points, metal will become anodic. These variations in concentration arise unhampered while ever corrosion continues.

The local variations are of no great size and by themselves would not cause heavy corrosive attack; but such attack as they do produce gives solid films of corrosion products such as rust. These are rarely uniform in distribution and are usually porous, so here and there will be cracks in the otherwise solid film. Through these and through the pores, oxygen from the solution passes to penetrate the metal and cause fresh corrosion. Meantime from these same regions metallic ions produced by corrosion are carried off more rapidly than those protected to some degree by the solid film. Corrosion is enhanced and may become dangerous.

Effect of Zinc Coatings

Zinc coatings produce an uninterrupted covering that keeps corrosive solutions away from the base metal, and also protects by electrochemical action when the covering is discontinuous. When exposed to highly acidic atmospheric pollution, zinc produces a self-protecting film of reasonably impermeable corrosion products, so shielding the metal from additional attack.

Small zinc regions in contact with large areas of cathodic

METALLURGICAL FACTORS

metals should be avoided, but small areas of other metals in contact with large regions of zinc are comparatively safe. Where the component needing to be protected by cathodic means is not or cannot be zinc-coated, the control of corrosion can be achieved by making zinc a galvanic anode in a cathodic protection system.

ENVIRONMENTAL EFFECTS

Nitriding and Fatigue

Nitriding is beneficial in increasing fatigue strength only when the work is of reasonable cross-section. In thin parts it sets up warping and a degree of grain growth. It is also unsatisfactory when the core of a stressed member is so light that the process causes it to fracture. Its advantage appears when components are subjected to high cyclic stresses and are of adequate size (see Fig 6).

The treatment compresses the outer layer of the metal and stretches the core, whose tensile stress is low directly

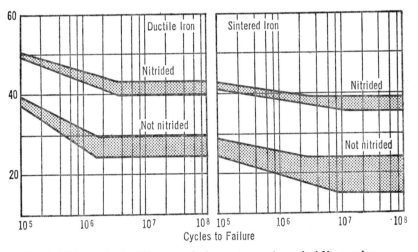

FIG 6 Effect of nitriding on fatigue strength: nitriding raises the fatigue strengths of ductile and sintered irons appreciably. Specimens of 3% copper sintered iron, ranging in density from 6.2 to 7.0 g/cm^3 were unnotched (1000psi = 6.895 MN/m^2).

below the nitrided surface, as in crankshafts. The compression stress of the core has to be balanced by the tensile load it supports; and the greater the core area, the smaller the unit stress required to maintain the layer of nitrite in compression. On the other hand, the smaller the cross section, the greater must be the tensile stress so that the compressive force of the nitrided layer balances it. This minimises fatigue failure by lessening the applied stress.

Induction Hardening

The effect of induction hardening on the fatigue life of a component is largely similar to that of case-carburizing, assuming the material and design to be identical; but on some parts, such as hydraulic transmission shafts, it gives considerably longer service life and better fatigue resistance by reason of the greater surface hardness obtainable and the residual compressive stresses in the case.

Hardness in itself has a linear relation with the fatigue limit of those hardened and tempered steels whose microstructure is almost wholly martensitic, up to Rockwell C35. With a hardness greater than this the fatigue limit increases to about 1.045 GN/m^2 (65 ton/in^2) at the same Rockwell figure. For example, a nickel chromium molybdenum steel of the alloy type when tested at 10^6 cycles showed that 50% of the testpieces failed and 50% withstood the stress. Fatigue limit is influenced by the heat treatment for a single hardness level in these steels, largely employed in the aircraft and automobile industries.

It must be emphasized that high fatigue properties in all these instances are dependent on and variable with the particular environment in which the component operates. It is therefore not possible to give an accurate forecast of specific behaviour unless the previous performances of the component under the particular conditions can be referred to.

METALLURGICAL FACTORS 49

Effect of Environment

Of late, a revival of interest in the effect of environment on fatigue has been observed. For example, the American National Bureau of Standards has carried out a large number of researches into this factor. Rotating beam tests in controlled environments have shown that for free-cutting brass, magnesium alloy AZ61A and steel AISI 4340, as well as Ti-Al and other alloys, fatigue strengths are much reduced by a moist atmosphere, the difference being in some instances as much as 39% less as compared to the same materials in a dry atmosphere. Less variation is shown, however, by brass. Moisture is an important factor as always in oxidation, and this may account to some extent for the decline in fatigue strength.

Some testpieces of metal, melted in vacuum, developed a fatigue crack at a level below the surface. The way in which any particular testpiece behaved when tested was seen to be influenced by its dimensions at the point of origin of the crack. Testpieces of magnesium alloy have given better results when coated with dodecyl alcohol. A staining effect was observed on the surface areas fatigued when magnesium was studied. Incidentally, the testpiece of this metal was notched, whereas the others were smooth, and all had straight seams except for the magnesium alloy, whose stem was tapered.

Effect of Nitempering

Nitempering is a recent product of American research and is a low temperature, case-carburizing process specially for strengthening ferrous metal components against wear and fatigue. The metal is heated to a temperature of 590°C (1,100°F) which renders it resistant to wear, galling, seizing and corrosive deterioration.

A second diffused layer is claimed to improve fatigue resistance and endurance. Low carbon steels are said to give longer service lives, the improvement being as much as

500%. In the same way, the service life of ductile iron is increased by up to about 80%, and that of malleable iron by about 25%. The thicknesses of the layer are governed by the length of time the components spend in the nitempering chamber of the furnace, the temperature to which they are heated, and the composition of each particular metal.

Effect of Metallic and Nonmetallic Contents

In an earlier section, nonmetallic inclusions and their effect on fatigue properties have been mentioned. There are, however, certain elements either inherent in the composition of such a metal as steel, for example, or added to give the steel particular properties.

Chief among these are, for example, carbon, lead and sulphur. Carbon content increases the fatigue limit when the hardness of the metal is consistently above Rockwell C45. Lead, which is introduced to increase the 'machinability' and for other reasons, does not appear to exert any noteworthy influence on alloy steels which have to withstand torsion fatigue; but a leaded carbon steel for shafts often shows so low a fatigue limit that it is not usually recommended for this purpose. The variation in results may, perhaps, be caused by the varying condition of the microstructure. Splined shafts are more sensitive to declining values of fatigue limit when their surfaces contain nonmetallic inclusions, and this makes the finishing of these surfaces important, so that it should be as high as practicable with commercial economy. The same remarks applying to the lead apply also to steels containing sulphur to improve the 'machinability' of a carbon steel.

Effect of Directional Flow of Metal

Some manufacturing operations such as forging, for example, give the metal a type of grain or 'flow' in a particular direction. Flow may, for example, be perpendicular to the load stress. Such a perpendicular directional flow does not increase the resistance of the metal to fatigue. Directional flow produces a concentration of stress of highly variable type. In consequence it appears impracticable to predict the fatigue resistance in these conditions for any particular component. There is, however, some relation between direction of metal flow and direction of load stress. Mechanical working of the surface of a metal renders all steels more resistant to fatigue; but at the present time no comparison in this respect of one steel with another is possible.

On the other hand, it is established that the fatigue strength of wrought stainless steel is adversely affected by free ferrite, one of its microstructural constituents, if it is rolled transversely to the direction of metal flow.

Effect of Hardness Gradient

In small testpieces of minor cross sectional area, the fractures normally originate at the surface subjected to the maximum stress. This is because surface irregularities cause concentration of stress. The test-pieces are hardened right through, but many components likely to undergo fatigue have a wide range of hardness from surface to core by reason of their greater mass and cross section. This is particularly true of case-carburized steels, in which the depth of case notably affects the fatigue limit. The precise effect is largely governed by the conditions of use, but it is not considered that fatigue strength differs greatly among steels of low or high 'hardenability'. On the other hand, a greater depth of case in certain carburized nickel chromium molybdenum steels embodying also a small vanadium content is advantageous for fatigue strength, while residual stress may also modify the test results as well as the service results.

Effect of Low Temperature

Temperatures below that of a normal room quickly increase the fatigue strength of a steel down to a temperature of $-130°C$ ($-200°F$). For temperatures above that of a normal room, plain carbon steels show a slight increase in fatigue strength at the outset, but some alloys of extremely low alloy content fail to show this increase. Constructional steels for sub-zero use are seldom chosen for their fatigue properties, but rather for their mechanical and physical properties.

Temperature also greatly influences the growth of fatigue cracks, which are less rapid in their progress towards a particular level, at lower temperatures. Despite the fact that longer cracks develop before failure takes place at higher temperatures, the cycles needed to produce failure are notably greater at lower temperatures. It is perceptible, therefore, that fatigue progresses much more rapidly at the higher temperatures.

Effect of Vibrational Stresses

As a result of recent investigations into the not uncommon failures of metal aircraft components through fatigue, it appears that sheet metals have to support vibrational loads, which are the basic cause of uneven loading in both tension and compression. The work now being carried out aims to establish a factor based on quantitative levels of surface stress and the extent to which component performance is governed by vibrational stresses. It is sometimes erroneously assumed that the fractures observed are laminations, but in fact the straight line markings running parallel to the rolling direction on the surface of a fracture of this type are the result of fatigue.

Effect of Soft Skin

Soft skin or decarburization has been discussed on page 40, and is the elimination of carbon from the surface of a steel by heating it in an atmosphere, the gases of which combine with the surface carbon to an extent above a certain value. The degree of decarburization decreases from surface to core, and at the surface the carbon may have been

Page 53 (above) Fractured face of shaft showing the markings of a typical fatigue failure; *(right)* 4340 high tensile steel component, showing fatigue from the corner

Page 54 Amsler electrohydraulic machine: static 100 tons fatigue ± 60 to bolted specimen; cabinet on left is the load programmer, controlled by punch tape demanding sequences of loads

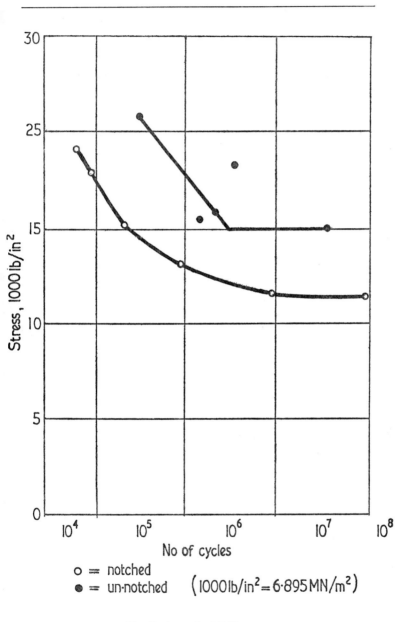

FIG 7 A typical S-N curve

completely lost. This loss of carbon at the surface is important in fatigue resistance, while surface irregularity is also a cause of decline in fatigue resistance. The soft skin and the irregularities act in a similar manner to notches. It is thus most important to prevent this type of surface imperfection. It is possible that some of the harmful effects of decarburization are remediable to a degree by recarburizing the material, or replacing the carbon by various methods. Another possible remedy is shot peening, the workhardening of the metal surface by the impingement upon it of a jet of metal shot under pressure.

Effect of Liquid Metals

Until recently, little was known of the fatigue behaviour of metals brought into contact with liquid metals. Recent research in Britain has dealt with the effect of sodium at a temperature of 500°C (930°F). This was found to be adverse in relation to the endurance limit of the steels. It appears that fatigue cracks are less numerous than those of low carbon steels not brought into contact with liquid sodium. As against this, however, they are more dangerous in action. Sodium does not appear to produce soft skin on plain carbon steel by a decarburizing action, and the S-N curves (see Fig 7) are not influenced by this liquid metal even though the secondary cracks are less numerous.

Copper in contact with mercury has also been the subject of experiment, and it is stated that the endurance limit of this metal based on 10^7 cycles is not affected. However, it is a different story with zinc, which has a notably harmful effect on copper, whose endurance limit at 10^7 cycles is lower. 70-30 brass in contact with mercury shows an intergranular fracture path. The effect of zinc on copper is heightened as the tin content of the brass embodying these elements increases.

Three per cent leaded brass is reduced in fatigue strength by contact with liquid metal. Lead resembles mercury to

METALLURGICAL FACTORS

some extent in its working effect on this type of brass, since at higher temperatures it becomes more active and slightly lowers the endurance limit based on 10^7 cycles. The S-N curves obtained are similar to those for mercury.

Stainless steel of the American 347 type is not greatly affected in fatigue strength by contact with liquid sodium at up to 500°C (930°F). Plain carbon steels, on the other hand, are reduced in endurance limit with increase of carbon content. In every instance the test data are modified by the temperature of the metal.

Effect of Testpiece Thickness on Fatigue Results

It will be appreciated that any factor affecting the results of fatigue tests is of considerable importance. It has been considered for some time that the thickness of a testpiece might have a marked influence on the growth of a fatigue crack. Up to the time of writing the complete effect is not yet established, but the current opinion is that variation in testpiece thickness is capable of modifying the loading stress condition. In practice the dimensions of the testpiece are important because they may have an effect on the development of a fatigue crack. Nevertheless, it is plain that much more work will have to be carried out as the data thus far obtainable are in some instances contradictory.

Effect of Vacuum Heating

The increase in the use of vacuum melting for those metals requiring to be exceptionally pure to eliminate as far as possible all undesired nonmetallic inclusions, has led to experiments concerning the advantage or otherwise of the process in extending the service life of the component subjected to fatigue. It appears that in the main, steels so heated are more resistant to fatigue than are those melted in the electric arc furnace. Nevertheless, this is not entirely true, because some components when melted or heated in the electric furnace give better results than do those melted or heated by vacuum. Research is still being carried out to ascertain, if possible, the reason why vacuum heating greatly lessens the incidence of short service life in testpieces. The hypothesis has been suggested that the presence in the metal of nonmetallic inclusions possessing the property of heat resistance is not sufficient to account for this phenomenon, so that there must be a more fundamental explanation, which has yet to be discovered.

Effect of Welding

Tests have revealed that constructional steels have their fatigue strength for every ratio of stress reduced to some extent by welding. Minor modifications of composition do not notably affect fatigue properties in these metals. If welding is essential, removal of weld reinforcement appears to give the metal a fatigue life almost equal to that of an unwelded metal, especially when the steel has to be used in constructions subjected to complete reversal of stress, which much improves the fatigue life. Bad finish lowers the fatigue strength in these steels, but this can be remedied by giving them a superior surface finish.

Effect of Annealing

The form of heat-treatment known as annealing in which a steel is heated and maintained at a suitable temperature, then cooled at a suitable rate so as to soften it and put it into the best condition for machining and cold working, is also beneficial in its effect on fatigued steel, whose service life under conditions of stress it prolongs.

CHAPTER 3
METALS FOR FATIGUE RESISTANCE

A feature of fatigue failure is that it does not produce a high degree of deformation in a steel component. The fact is that when a metal has been subjected to repeated loads of the same type and in many cases continually reversing, and fails as a result, its fractured surface is greatly different in appearance from that of a metal whose fracture is caused by the too heavy applied load. The principal factors in the fatigue failure of steels include the method of manufacture, the method of working, whether or not welding has been carried out on it, any heat treatment to which it has been subjected, the type of surface treatment, the finishing processes, and lastly, the conditions under which the metal will be employed.

Choice of any particular steel for fatigue service is largely a matter of compromise in which the high fatigue resistance of the material is weighed in the balance with other factors, such as ductility, sensitiveness to notch effects and other surface blemishes, the high ratio of yield strength to tensile strength, the impact resistance of the steel, its suitability for the conditions of use, its ability to be welded or formed, and so forth.

A steel called upon to undergo 6×10^6 alterations of equal tensile and compression stress has on average a fatigue limit of about 45% of the ultimate tensile strength, which is the regular method of determining fatigue strength in metals. This strength is, however, governed by the manner in which the stress is applied, but the essential factor is the stress range. (It is most important to note, however, that *the majority of metals do not show a fatigue limit*.) With metals other than steel, the results obtained by the above method are not so closely related to ultimate tensile strength; some of the lighter alloys varying considerably in this respect owing to the way in which they have been manufactured or worked. Fluctuating values are obtained from brasses and bronzes in particular.

Carbon Steels

Low carbon martensitic steels (0.08-0.15% carbon) have a fatigue limit of 35-50% that of the tensile strength 900-1,300 MN/m^2 (58-85 tons/in^2). These steels are being used for crash bars, welded tubes, fasteners, small spring-type components and corrugated panels, as an alternative to aluminium, titanium, stainless and maraging steels, and nonmetallic materials.

Carbon Hot-finished Steel

A direct relation is found between the tensile strength and surface condition of this steel and its fatigue properties. The machined steel has a fatigue limit of about 40% of the tensile strength, but if the fatigue stresses are concentrated at the surface of a forged or hot-rolled steel, the values will be greatly reduced owing to soft skin, superficial coarseness and surface defects. Hence, it is always advisable to machine off the asperities on those surfaces where critical stress will be encountered. On the other hand, an allowance can be made for the inferior strength of the hot-finished surface.

In these steels, nonmetallic inclusions adversely affect the fatigue limit in most instances, and if of large size are harmful when the component has to withstand impact or fatigue. 'Directionality' (see p50) is important also, because when steel is hot-rolled, the inclusions are elongated in the direction of rolling under plastic flow. This is true also of chemical segregations—impurities non-uniformly distributed and governed by the chemical composition of the steel and the rate at which it cools. These occur during the solidification of the steel from the melting temperature. The main effect is on ductility and resistance to shock, but they are not notably harmful to the steel's strength.

Seams are the most injurious surface blemishes wherever fatigue loading is involved. Decarburization or soft skin renders the steel unsuitable for fatigue resistance, and should be machined or ground off. Cyclic loads need the maximum resistance giving 10^6 cycles of stress as a life expectancy. At this rate a closed die steel forging should give a minimum permissible stress of about 150 MN/m^2 (9.75 tons/in^2) when not machined, but after machining, the value rises to about 310 MN/m^2 (20.5 tons/in^2). These values relate to completely reversed stress cycles for a steel of 269-285 Brinell hardness, and the difference between the two values indicates the effect of a decarburized surface and one produced by forging when compared to a surface of the same material machined and ground.

Constructional Steels

The fatigue strength of constructional steels is not covered by standards, but must be considered as related to the specific steel. Suitable rules for designers have been drawn up in the United States under the classification As CA1. High strength steels appear to show properties closely similar to those of low carbon steel.

Some steels of 0.2% maximum carbon content are being used in modern drilling platforms for oil rigs. These are subjected to repeated loads, and have to possess good fatigue properties as otherwise the fluctuating stress produces failure from fatigue at strengths below what is permissible with static loads. These steels can be readily welded to each other and to hot-rolled heat-treated steels. They are of heat-treated carbon steel of high strength, and have a yield strength of about 340-700 MN/m^2 (22-45 tons/in^2). Alloy steels of similar carbon content and type have also been recently developed.

The constructional steels must withstand high working stresses and have a great flexibility when being used on site, so that their fatigue properties are highly significant. Incidentally, tough fracture-resistant steels are being developed by BISRA (British Iron & Steel Research Association).

Matrix Steels

These are much used today for components needing great toughness, strength and fatigue resistance. They are claimed to have a more effective fatigue strength than high speed steels owing to their secondary hardening property. This means that they are given an additional hardening by tempering after quenching; the austenitic microstructure being changed into the harder martensitic. Steels of this class have a typical composition of 0.5% carbon, 2.0% tungsten, 4.5% chromium, 2.75% molybdenum; or 0.55% carbon, 1.0% tungsten, 4.0% chromium, 1.0% vanadium, 5.0% molybdenum, 8.0% cobalt. They are usually employed for punches, cold heading tools, gear rolling operations, extrusion tools and fasteners, and also for torsion bars, gears, shafts, bearings, die castings and thread rolling dies.

The steels show a higher fatigue strength at a lower number of stress cycles, and as the strength increases, so the stronger steels show a higher fatigue stress. When joints of these steels are welded it is possible to obtain extremely great advantage from removal of the weld reinforcement, which strengthens the joints against fatigue.

Steels for Cold Drawing

Cold drawing affects the microstructure of steel by increasing the fatigue limit, among other properties, to a value higher than that of steel in the hot-rolled condition. Fatigue limit is also raised in these steels by a form of heat treatment known as stress relief; but in both instances the results depend on the degree of cold working applied. The latest research suggests that higher austenite content of the steel causes a decline in fatigue limit amounting to as much as 10% on occasion. When cold-worked steels are being produced, the flow of the steel is first directed in hot working so that the grain runs in what is known to be the direction to suit the eventual form of the finished part and its function. This in itself improves resistance to fatigue, and the cold working operations that follow in making the component do not change the fatigue resistance adversely, even though hot working has come to an end some time previously.

It should also be noted that these steels, when of carbon type, can be austempered rather than normally heat treated. Austempering is an interrupted quenching operation in which the steel is raised to a suitable temperature above the critical range to render its microstructure austenitic; but it is not then cooled to room temperature in a quenching medium. Instead, it is transferred to a hot quenching bath held at a predetermined constant temperature below the critical range, but above the martensitic change point. It is then maintained at this temperature long enough to ensure that the austenite is transformed into the final microstructural constituents, after which it is cooled to room temperature.

Carbon steel components, so treated, have shown higher fatigue and impact resistance in such parts as the blades of harrows.

METALS FOR FATIGUE RESISTANCE

Wrought Steel

In general, wrought steel has a higher fatigue limit than notch-free cast steel, but if both wrought and cast steels are notched or left rough on the surface, the fatigue limits of the two metals do not greatly differ as regards either tensile strength or value. The tensile strength is in the region of about 1,850 MN/m^2 (120 tons/in^2).

Cast Steels

As with other metals, cast steels or steel castings are greatly reduced in fatigue strength by surface defects, or by being left in the as-cast surface condition. When the cast parts are machined on their surfaces, the fatigue limit is about 50% of the tensile strength for a plain carbon manganese steel of composition ranging from 0.19-0.30% carbon, 0.68-1.19% manganese.

Alloy Steels

The range of alloy steels is wide, and it is impractical to detail here the fatigue properties of every one of them. A few have been picked out because of their popularity in fatigue resistance, or because they are in themselves interesting in relation to fatigue. One of these is a steel containing a modicum of silicon and nitrogen. This has not been extensively used for fatigue properties, and tests have so far been minimal; but such figures as are available show it to possess a fatigue ratio to tensile strength of 0.55, and it has therefore good fatigue properties combined with good service life under the conditions. It is being used for brackets, wheel rims, seat track rails of automobiles, reinforcement beams, etc, and in particular for the base-plates of office swivel chairs.

A steel with 1.25% manganese, 0.22% carbon, 0.30% silicon, 0.02% copper and 0.02% vanadium, is of high strength and is intended for fatigue and corrosion resistance combined. Its tensile strength range according to its thickness

METALS FOR FATIGUE RESISTANCE

is 460-480 MN/m² (30-31.5 tons/in²), and its endurance limit in the as-rolled state averages 290 MN/m² (19 tons/in²). Another high strength steel contains 0.16% carbon, 0.75% manganese, 0.02% columbium (niobium), and in sheet form has a tensile strength range of 430-580 MN/m² (28-38 tons/in²) and an endurance limit of 55-60% of the tensile strength. These steels are used for earth-moving equipment, shovel blades, constructional members for building, railway trucks, highway plant, gas cylinders, hot water cisterns, agricultural plant, tractor plates, cement road forms, industrial containers, etc.

The fatigue strength of alloyed carbon steels is greatly heightened by the introduction of the alloys when the steels are of low alloy high strength type, with the alloying elements present in minor quantities only. Constructional steels, quenched and tempered to render them suitable for the greatest possible load carrying capacity, have a composition range of 0.12-0.21% carbon, 0.4-1.0% manganese, 0.15-0.9% silicon, 0.2-0.4% chromium, 0.15-0.65% molybdenum, 0.0005-0.0025% (max) boron. On occasion, small amounts of alloying elements such as titanium, copper, nickel or zirconium are introduced. The ordinary curves used for presenting facts about metal fatigue are less valuable to designers than constant life diagrams.

Extra-fine grain microstructure improves the fatigue resistance of steels containing 0.18% carbon, 1.50% manganese, 0.06% silicon, 0.005% vanadium, 0.02% aluminium and 0.005% nitrogen. These are used because they show improved fatigue resistance when used for truck trailers, railway equipment and heavy construction machinery. Plates of these steels have fatigue limits above 270 MN/m² (17.5 tons/in²) in reverse bending, these limits being governed by the inherent strength of the steel and the condition of its surface, which shows virtually no soft skin. The fatigue limit given as 5-8% is obtained at 10^7 cycles, and assuming the surface is polished, about 380 MN/m² (24.5

tons/in²) is the reverse bending fatigue limit.

Irons

Grey irons have fatigue limits governed by the type of reversed bending stresses to which they are exposed, and are affected by both temperature and tensile strength (see Fig 8). In a notch free surface condition, the fatigue limit for an iron of 2.84% carbon, 1.12% silicon, 1.05% manganese, is about 139 MN/m² (9 tons/in²) whereas for a notched surface it is about 120 MN/m² (7.75 tons/in²) on the gross

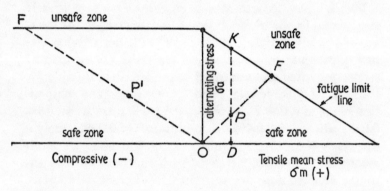

FIG 8 Stress range for cast iron under ranges of repeated axial stress superimposed on a mean stress. P=conditions of tensile mean stress; ratio of OF to OP or OF' to OP'=safety factor; DK/DP=safety factor for constant mean tensile stress conditions

area and 145 MN/m² (9.4 tons/in²) on the net area when the applied stress is about 170 MN/m² (11 tons/in²) for the notch free iron, and 185 MN/m² (12 tons/in²) for the notched. In these conditions the unnotched iron sustains about 10^8 cycles as against the notched iron's $10^{7.6}$, at which point failure occurs. The fatigue limit for this iron does not show so wide a variation in the three types of results. These figures are for complete reversed cycles of bending stress. High temperature greatly reduces fatigue strength, and

METALS FOR FATIGUE RESISTANCE

tensile strength beyond about 370°C (700°F).

Not all the loads of loading cycles imposed upon grey iron are fully reversed, that is, the axial or torsional loading is often not complete in this way. Since it is costly to run tests of fatigue limit in every instance, a simple working rule can be followed when some notion of the limit is necessary for design. This is: $\frac{T \times 35}{100} = L$, where T is the minimum specified tensile strength and L the reversed bending fatigue limit. This is a reasonably trustworthy value for machined grey iron castings.

The fatigue limit for all mean stress values is represented by a curve that gives the range of stress on the fatigue limit as nearly as possible, with an approximation for the maximum value of alternating stress. If the real stress cycle is known, this chart can be used to determine an approximate safety factor, represented by P in the diagram and established by the line showing the distance of OF^1/OP^1. There are certain limitations of this diagram (Fig 8) however, as it cannot be applied to systems in which the mean stress is constant but mechanical vibration occurs. Here the safety factor is represented by the link DK/DP.

Under conditions of fatigue with irons of relatively low strength, there is no great diminution of strength caused by sharp changes in cross section or notched surfaces; but holes and fillets lose some small amount of strength for these materials. Irons of greater strength lose much more fatigue strength by the presence of these faults of design or manufacture.

Nodular iron has a fatigue limit in the notched condition of about 127 MN/m² (8.25 tons/in²). For the notch-free state the limit is about 210 MN/m² (13.5 tons/in²) for a stress concentration factor of 1.67 and a fatigue ratio of 0.43. This corresponds to a tensile strength of about 480 MN/m² (31.25 tons/in²). For a pearlitic nodular iron the corresponding fatigue limits are about 278 MN/m² (18 tons/in²) for the notch free, and 370 MN/m² (24 tons/in²) for the notched material, stress concentration being the same and the fatigue ratio 0.40. The tensile strength in this instance is about 680 MN/m² (44 tons/in²). Polished stainless steel components have a higher fatigue limit than oxidized parts with the typical surfaces of heat treated castings.

Austenitic Manganese Steel

Commonly known as Hadfield manganese steel, with 12-14% manganese, this steel has a fatigue limit of about 260 MN/m² (17 tons/in²), or about 0.34 times the tensile strength.

Heat Resisting Alloys

Varying widely in composition and properties, these steels display fatigue in differing degrees of strength and characteristics. Martensitic stainless steels containing 11.5-12.35% chromium, 0.15% carbon, 1.0% manganese, 1.0% silicon, 0.25% nitrogen, are sometimes called upon to withstand an axial alternating force up to ±35 MN/m² (2.25 tons/in²) when mechanically oscillated. Another stainless metal of ferrous base containing nickel, chromium and molybdenum has to meet superimposed alternating loads at four different temperatures in some operations, the stress being combined axial and bending. Axial fatigue is experienced by cobalt alloys having a nickel base. Each heat resisting steel must be tested to fatigue strength to determine its endurance limit, the unit stress below which an indefinitely large number of cycles can be withstood without failure. For most steels this represents from 2/5 to 3/8 of the ultimate tensile strength.

Stainless Steels

As these, like many heat resisting steels, are of varied compositions, it is not practicable to list the fatigue limits for each type. In general, they are influenced in both strength and endurance limit by 'directionality'—that is, the relation of the axis to the precise direction of the metal flow, usually determined by the position of the significant crystal planes in relation to any fixed planar system, and converted into a crystal deposition parallel to the direction of deformation. Other influential factors are temperature, surface condition and microstructure. Polished surfaces are much superior to those produced by immersion in a pickling bath or those covered with scale when they emerge from the production processes. The surface affects the steels increasingly as hardness declines. Fatigue limits for the steels lie between 235 and 1,170 MN/m^2 (15.25 and 76 $tons/in^2$) approximately, according to the type of steel, its condition, method and production and hardness.

Temperature affects the fatigue strength and limit of martensitic stainless steels for use in turbines, etc., but the alloy content modifies the range of values. Surface condition also varies the fatigue properties on occasion to a considerable extent, and as in so many instances, the higher the polish, the greater the improvement over rough or pickled surfaces. The greater the hardness, the greater the response of some steels of this kind to change of surface condition. In most instances the variation is governed by the manner in which the hardness is obtained, that is by cold working or heating for treatment.

Unless differences in composition are considerable, they will not have a marked influence on fatigue at room temperature as long as the hardness remains unchanged. A change in hardness at room temperature is encountered with those steels liable to age. Ageing is a process of microstructural change slowly taking place in those metals susceptible to it, and in alloys at atmospheric temperature. As a result, the proof stress, maximum stress and hardness values increase, but ductility is slightly reduced. The cause of these effects is usually a submicroscopic precipitate from a supersaturated solid solution. At higher temperatures the ageing effect is produced more rapidly.

A microstructure containing free ferrite is, however, injurious to the fatigue properties of these steels.

18-8 chromium nickel and other austenitic stainless steels, cold worked and in fully stable or unstable conditions, reveal a gain in fatigue strength when in alternating stress, a gain disclosed by the effect of cold working on the fatigue limit. 25-20 chromium nickel alloys and those containing 0.6% carbon, 7.5% manganese, 8.5% nickel, 4% chromium, are improved when in the stable austenitic state, but the improvement here is relatively small. Martensite in the microstructure of the non-austenitic stainless steels does not appear to show an improved fatigue limit for the steels when cold worked.

The degree to which the steels are cold worked does not in itself account for the improvement shown elsewhere, and it is reasonable to believe that some factor at present unidentified must be found before it can be satisfactorily explained.

Generally, these steels show better fatigue resistance with increasing tensile strength and hardness, while the temperature to which they are subjected lowers both fatigue strength and fatigue limit in certain martensitic stainless steels. At normal temperatures, equal hardness of the steels and similar composition mean equal fatigue properties at normal temperatures. If, however, free ferrite in the microstructure is high the fatigue resistance is considerably reduced.

Nonferrous Metals

Some interesting studies have been carried out in relation to the fatigue of metal crystals, especially those of silver and zinc. From the data obtained, it appears that the fatigue process is possibly made up of a primary stage in which hardness increases with dislocation density, that is, the possible vacant atom spaces in metal crystals and their

age 71 (above) Schenk machine: resonant machine 20 tons. Modified to be cited by electrodynamic shaker (on left) in place of rotating mass, enabling the plication of variable amplitude loading; *(below)* Dowty electrohydraulic achine, 20 tons, with control cabinet and block programme facility on the right

Page 72 (left) Fracture surface of a high strength steel (En 24) tested in push-pull fatigue at a stress of 50 ton f/in². The fracture was initiated at the surface of the specimen (bright region) and later propagated on planes both normal and inclined to the direction of loading until final test fracture occurred *(below)* Electron-fractograph of fatigue striations in a high-tensile steel (En 24). The striations are irregular and are now well defined which is characteristic of these low-alloy martensitic steels. The direction of crack propagation is from the low right-hand corner to the top left hand corner of the fractograph

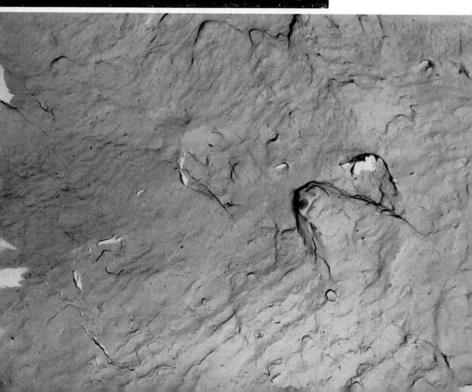

METALS FOR FATIGUE RESISTANCE

closeness to one another. These vacant spaces may shift when stressed, and some workers in this field believe that these shifts explain plastic deformation, creep, diffusion and other phenomena. A secondary stage of greatly decreasing integrated intensity and recovery is then followed by a third and final stage typified by considerable fragmentation of the microstructure. Apart from a higher rate, high stress fatigue appears to be in principle identical with low stress fatigue. A failure of fatigue type occurs, it is believed, as a consequence of high dislocation density lattice regions acting as an internal obstruction to cross slip.

Sintered Carbides

Sintered carbides of metallic type are hard and brittle, and some of them, chiefly tungsten, titanium, tantalum and boron carbides, are put to many different uses in manufacture. Mostly the tungsten and titanium carbides are used to tip cutting tools for work on metal machining. These carbides have a fatigue limit governed by the composition, but it is difficult to draw conclusions of importance from the data available. The fatigue limit for the metallic carbides ranges from 580-720 MN/m^2, the maximum value being for a carbide containing 13% cobalt as a binder. These figures are based on 20×10^6 cycles.

Carbides used for cold extrusion dies normally include 10-25% cobalt in their composition, and have to show great fatigue resistance when subjected to the cyclic stresses encountered in these operations. They are also resistant to heat and corrosion, being of sintered composition, the hard titanium particles of carbide being held in an alloy matrix. They are virtually free from fatigue, and in addition are also largely free from shock, wear and galling, as well as heat shock.

Plated Metals

These are metals to which a film has been applied on

the surface by an electrochemical process. These metals are likely to undergo internal stresses varying in proportion to the thickness of the deposit. The stress produced may be severe enough to generate a pattern of cracks in the plating caused by the contraction of expansion of the metal, consequent upon deposition, and leading to distortion. Heat-induced behaviour of this kind is invariably associated with lowered fatigue properties in the base metals when plated. The remedy is a measure of compressive stress, accomplished by using a plating solution containing nickel sulphamate.

Zinc Alloy Die Castings

These castings have a fatigue limit of about 180 MN/m^2, and wherever used under conditions of cyclic loading, this should be taken as the operating stress which must not be exceeded. However, it is not usual for these alloys to demand high strength under constant load.

Titanium and Titanium Alloys

Rocketry, space applications, and developments in other directions have vastly increased the importance of titanium and its alloys. The disaster to the Comet aircraft whose destruction was primarily caused by metal failure from fatigue, demonstrated beyond contradiction that the most serious risk confronting any aircraft is fatigue of the metals employed in its construction. The use of titanium and its alloys in various applications has, therefore, made the fatigue properties of these metals for high temperature service in aircraft, of the highest importance.

At present not enough is known on this subject, since the number of tests carried out have not been sufficient to assure designers and engineers of the response of these metals to every type and degree of fatigue stress. Consequently, the practical steps open to the manufacturer are in testing each component when complete, and in working

METALS FOR FATIGUE RESISTANCE

out a design that ensures maximum soundness of structure, with provision for any potential failure. In other words, each construction and component must have the properties necessary to ensure that failure will not cause the complete collapse of the machine. Moreover, when the aircraft goes into service it must be continually inspected and tested for fatigue effects, in particular local concentrations of stress, the frequency of the load cycle, the cross sectional dimension, corrosion or the risk of corrosion, residual stress, the load frequency range, the microstructural condition of the metal and the state of the surface.

For commercially pure titanium, the fatigue limit at 10^6 cycles is about 270 MN/m^2 (17.75 tons/in^2) when the tensile strength is between about 515 MN/m^2 and 615 MN/m^2 (33.5 and 40 tons/in^2). The relation between fatigue limit and tensile strength for alloys of titanium is not readily comparable with that of hardened and tempered steels, as the results of tests vary.

Joints and fastening devices of titanium alloys have to possess maximum fatigue resistance. The thickness of the sheet is more important than the precise composition in this respect. Another important factor is the design of joints. The joint fatigue strength is particularly affected by both factors. As regards composition, the constructional joint will sometimes derive much more improvement from an alloy of different composition having higher static strength, than from a better design of joint. For these particular features of an aircraft, a higher fatigue strength is shown by alloys of aluminium.

The fatigue strength of the titanium alloys has been studied in some detail. For example, the cyclic impact load under the impact strength of particular alloys has been tested for such titanium alloys as alpha-beta Ti-6A-8U-5 Fe and beta types Ti-13V-11Cr-3Al. Broadly, the alpha-beta types show up more favourably than the beta types under this kind of loading. At 10^7 cycles the limit is just over 480

MN/m^2 (31 tons/in^2) for the first of these alloys and 340 MN/m^2 (22 tons/in^2) for the second. The tests covered long cycle fatigue ($10^4 - 10^7$) of normally procurable titanium alloys. Emerging from the tests of impact fatigue was also the discovery that crack initiation and growth are active causes of fatigue, but this activity declines as the alloys increase in tensile strength, failure from impact fatigue being thereby lessened. The immediate problem is said to be the low wearing power of the alloy Ti-6A-14V when subjected to impact fatigue.

Aluminium Die Castings

These are formed with dies of steel likely to fail because of surface cracks. The surface layers alternately contract and expand until the steel attains its fatigue point, at which it becomes unable to withstand the stresses imposed upon it. Corrosion and erosion also attack this layer, while a mechanical splitting action is consequent upon the forcing of alloy fins under pressure into the cracks. The moment the crack begins, the next batch of alloy in the die at once fills it with molten metal, which, on solidification, remains in the crevice. Each time the die is put to work this wedge of solid metal is hammered deeper and deeper in, so that the crack is steadily extended until fatigue failure occurs. Hence, dies for this work have to be made of a highly alloyed metal.

Copper Alloys

Among the nonferrous alloys, those of copper have often to withstand severe fatigue, and for this reason are usually made of superior metals for such parts as springs, diaphragms, bellows, flexible hose, etc. The most suitable alloys in order of increasing strength are silicon bronze, nickel silver, A, C and B phosphor bronze, and beryllium copper. Fatigue limits are governed by the condition of the surface and the temper of the alloy, but in addition, the extent of any corrosion likely to be encountered is also important. Table II gives the average values of fatigue limit for these materials, approximately.

TABLE 2

Material in Strip Form	MN/m²	Tons/in²
Silicon Bronze A	158	10.25
Phosphor Bronze A	154	10.0
Nickel Silver B	155	10.10
Phosphor Bronze C	188	12.25
Beryllium Copper	250	16.25

All the above materials are provided with spring temper

Cartridge Brass

Spring temper of this type of brass does not preclude the possibility of failure from fatigue, because the alloy has no tin content, and this may lead to disastrous results. Admiralty brass containing 1% tin increases the fatigue limit while still remaining in the spring tempered condition, and is capable of withstanding up to 25% more cycles, so eliminating failure in certain applications.

Aluminium Alloys

Aluminium alloys depend for their fatigue strength to a considerable extent on the chemical composition of the alloy and the orientation of the grains or crystals, but the degree to which the metal is cold worked or hot worked also plays a part. There is no great difference in fatigue strength between aluminium plates, extrusions and forged parts, nor is there any practical significance in the direction of working or the design of the component unless these are extremely unsuitable for the conditions of use.

Powdered Metals

Sintered steels with 0.48% carbon and 4% nickel are said to have fatigue limits rising with density, and alloy contents within the ranges 6.6-7.2 g/cm, and 0.003-0.08% carbon, 2.0-7.0% nickel. Definite fatigue limits are found between 10^6 and 10^8 cycles. Heat treatments that increase tensile strength also increase fatigue limit.

Fatigue ratio remains at 0.4 through the strength range for smooth testpieces, but declines above 690 MN/m² (45 tons/in²) for notched testpieces.

Transmission turbine hubs of copper-infiltrated iron powders are periodically tested for resistance to fatigue. They are first loaded on to a fixture and anchored to a testing appliance. Cyclic torquing at 1,700 Nm (15,000 in/lb) is carried out and a minimum of 200,000 cycles is essential. If the hubs are not infiltrated, cycling is 10^6+ at 815 Nm (7,200 in/lb). The iron powder has a composition of 0.7% carbon, 3.0% copper, the balance being iron, and when pressed, sintered and sized, gives a density of 6.6×10^3 kg/m³ and a tensile strength of about 445 MN/m² (29 tons/in²).

CHAPTER 4

BEHAVIOUR OF PARTICULAR COMPONENTS

Sleeve Bearings

These show a sharp increase of fatigue limit when lead is added to a babbitt metal, but, if the steel back constitutes too thick a lining, the fatigue properties decline. This statement involves the assurance that temperature and pressure of use are constant. A bearing metal of this type is much more fatigue resistant than a babbitt metal of tin base, but the addition of lead to the tin base alloys merely reduces their fatigue strength, especially at the higher temperatures of service. An alloy containing both copper and lead has a much higher fatigue resistance than a lead base babbitt metal for the same type of bearing, and also a much greater load carrying ability. At constant temperature and load a plated overlay gives an alloy containing 35% lead a much lower load carrying capacity than one containing 24% lead.

Some bearings are plated with lead, tin and copper alloy to a depth of 25μm (0.001in) but the fatigue performance of these is much inferior to that of bearings plated with silver plus about 0.5% lead, which, however, are themselves inferior to those with a lead content increased to 4% with a lining thickness of 5.56mm (0.22in). These last have a much longer service life than silver plated bearings, while silver plate itself lasts longer if the thickness is cut down to 100μm (about 0.004in).

Broadly, tin base and lead base babbitt metals have about the same fatigue strength; but with thinner material the strength is nearly twice as high, the precise range for copper lead alloy being 50-150μm (0.002 – 0.005in). An identical thickness range for copper lead alloy more than doubles the fatigue strength. Sintered carbide and infiltrated bronze also increase the length of service given by the bearings, while a further improvement still is found with aluminium alloys of 150-450μm (0.005-0.015in) thickness; but for all these bearing metals the fatigue strength amounts to about 37,800 N (3.8 tons)/100 hr, the ability to carry a load remaining the same. The presence of a film of lubricant on the surface of bearings of this type changes the pattern of stress operating on a specified unit volume and therefore alters the fatigue life in service.

Engine Valves

Engine valves are subjected in service to alternating stresses involving fatigue, and the particular degree of strength required in any particular valve is governed by its function as part of the engine. No specific relation exists between the fatigue strength of a valve metal and the ability to perform its function, as demonstrated by the fact that some valves of low fatigue strength work satisfactorily. If, however, a number of valves of low fatigue strength continually break down, it is advantageous to have facts and figures of fatigue properties for specific materials at hand. A wide range of different metals and alloys for valves is available, some designed for high fatigue limit, others for particular ranges of stress. In general the stress range is from ± about 112 MN/m^2 (7.25 tons/in^2) for temperatures varying between 730 and 870°C (1,350-1,600°F) according

BEHAVIOUR OF PARTICULAR COMPONENTS

to the composition of the alloy. The fatigue limit range is from 10^6 to 10^8 cycles over the same range for the principal alloys covered by these data, some of which, such as Nimonic 90, are for a duration of only 300 hours.

The lives of engine valves are frequently abbreviated by thermal fatigue, and austenitic stainless steel valves appear to offer a much greater resistance to this than a hard brittle nonmagnetic steel of the sigma phase type. This is the phase in which a hard brittle nonmagnetic compound occurs. It seems that atoms of the compounds of the phase may be replaced to some extent by the atoms of other elements. Sigma is believed to cause serious embrittlement, especially when a metal has been exposed for long periods of time to a particular range of temperature.

The most effective engine valve alloy from this point of view is said to be a 21% chromium, 9% manganese, 4% nickel, with a small nitrogen content. A valve of this alloy has functioned without cracking of the head for 28 hours as compared to an alloy of 24% chromium, 5% nickel, 3% molybdenum.

Welded Joints

When these are made of aluminium alloys they may be required to withstand fatigue at high loads, and then their performance depends largely on the composition of the alloy welded. With decreasing load, the differences slowly disappear until at about 10^6-10^7 cycles, the fatigue strength of the alloy welded by the electric arc process is almost 50-70% that of the unwelded material, a percentage not affected by the composition of the alloy. Thus, three different aluminium alloys, with butt joints arc welded, are able to withstand nearly 10^8 cycles of stress without failure, despite having maximum stress differences of about 70 MN/m^2 (4.5 tons/in^2) between them.

Tubes

A considerable range of metal tubing is employed for

work involving much flexure, often of great severity. Typical examples include arcing horns on electrical switches, aircraft antennae, flexible hose and bellows, etc. The essential factor here is vibration, which must be absorbed, while the movement of other parts must also be resisted. Hence fatigue resistance is exceptionally important, being governed primarily by the properties of the material used. The tubes should have walls of usually small cross section, and the surfaces of the metal must be of the highest quality to prevent as far as possible fatigue cracks from developing.

Corrosion must also be prevented to the maximum degree because the pitting caused on the tube surfaces constitutes the focus of corrosion initiation and quickly produces fatigue failure, while if general and heavy corrosion occurs, it will completely ruin the material. The superficial flaws caused by faults in processing the material of the tubes also concentrate the starting points of fatigue cracks.

Resistance of tube metal to fatigue increases as the tensile strength increases, but a point is reached at which embrittlement occurs, and then the resistance to fatigue is nil. Aerospace demands have enhanced the development of precipitation hardening steels of stainless type for the flexible hose employed in and around jet and rocket engines because these can withstand high temperatures. Among the alloys for this function are an age hardening alloy known as Inconel, and a stainless alloy steel of precipitation hardening type possessing high fatigue strength combined with resistance to corrosion and high temperatures up to 980°C (1,800°F) for the lower stresses.

Improved piercing and extrusion practices have also enabled thin-walled flexible tubing to be made with virtually no unsightly and harmful seams, nonmetallic inclusions, laps—defects caused by part of the steel folding over on itself and therefore failing to be welded up in further rolling or forging. Other defects thus prevented are segregated areas. Cold drawing and annealing followed by welding are

BEHAVIOUR OF PARTICULAR COMPONENTS

said to produce tubes equally acceptable in performance with those of seamless metal.

Springs

Springs are often subjected to repeated cycles of stress under load, covering a range between the starting and final positive values. The difference between the starting and final stresses of the cycle for a spiral steel spring that will not fail during the cycle is commonly termed the range of shear stress. This range slowly declines with increase of the minimum and maximum loading cycle, the permissible maximum rising to a value at which the spring undergoes permanent set, that is, a degree of plastic deformation not remedied by the removal of the deforming stress.

Springs of the lighter class are normally manufactured from steel wire, and the smaller wire diameters apparently give the higher fatigue results. Torsion springs of piano wire, for example, have often to support a minimum cycle life of 50,000 cycles, but by no means all wires of this type fulfil this specification. The variations may arise from difference of tensile strength in the steel owing to the modification of wire diameter.

Many springs subjected to compression are, like most others, harmfully affected in fatigue properties by surface blemishes and flaws, so that fatigue life can be considerably increased by care during manufacture. Thus the spring wire requiring the finest possible polished surface must be ground before being coiled, but this is for many purposes uneconomical, so that it is a practice adopted only when the service makes it imperative. The extent of surface damage depends to a large extent on the quantity of steel removed during the various stages of production.

Seams, which are longitudinal surface defects in the form of shallow grooves or striations, cause much trouble; the degree depending partly on their location, being greater for those at the mean diameter and less for those on the outside diameter. On the other hand, springs with a large index suffer more from seams and other faults on their external diameters.

Pipes and Cast Components

When these are made of malleable iron, the iron is usually in the notchfree ferritic state with a fatigue limit of between 173-204 MN/m^2 (11.25-13.25 tons/in^2), that is, about half the tensile strength of the material. Fatigue strength declines in relation to the notch radius for ferritic iron, failure occurring at a notch depth of about 1.6mm (0.08in). Pearlitic malleable iron has a superior fatigue ratio when in the oil quenched and tempered condition, its fatigue limit lying between 30% and 40% of the tensile strength. These figures relate, of course, to pipes and other castings.

AIRCRAFT FASTENERS

Where, as in aircraft, fatigue life is critical, the type of

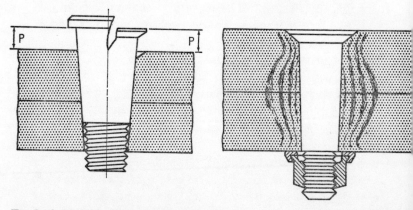

FIG 9 *(left)* Self-aligning interference fit bolts: the design of tapered interference-fit bolts makes them self-aligning and holds them in correct position for torquing or driving them the required distance (P). Heads may be flush or protruding; *(right)* preload stresses induced by tapered fasteners prevent normal service stresses from reaching the hole and causing failure

BEHAVIOUR OF PARTICULAR COMPONENTS

fastener termed the 'controlled interference-fit fastener' is often used. This is claimed to align itself and remain in proper position for torquing or driving to the desired extent. (See Fig 9.) The heads are flush or projecting. The mating hole is tapered so that the fastener is a close fit throughout the greater part of the hole. It is driven or pulled into full interference. The fastener has the effect of enlarging the hole into which it is inserted so that radial and peripheral tension stresses are generated around the entire circumference. Thus, preload stresses stop ordinary service stresses from reaching the hole and causing fatigue breakdowns, since the induced fatigue stresses in a construction have to be higher than the preload stresses before they can reach the hole.

Rivets

Interference-fit rivets are also being embodied in aircraft at present being built, and as the new materials being used for the construction are not susceptible to high preload stresses, considerable advantage is obtained by using the medium to high interference levels provided by these rivets. For preloading in both rivets and bolts of these types, the hole to be riveted should have a good finish, free from spiral grooving, or the galling produced by chips. It must not have a bell mouth, and is best produced by drilling and reaming at the proper feeds and speeds, and with the best lubricant, using an automatic or semi-automatic drilling machine. Steel, titanium alloys and stainless steels are used for these fasteners.

Parameters

Loading parameters in fatigue differ for the loading and testing of aircraft fasteners. A parameter is a line or figure indicating a particular point, line, figure or quantity in a class of quantities, lines, points or figures, constant in particular instances but variable in others. In service, a tension fastener first introduced into an aircraft has to meet a comparatively severe wrenching torque, not permitted in testing. In testing, moreover, exceptional precautions are taken to ensure rigid concentric loading, whereas in service, aircraft rarely attain this rigidity or concentricity of loading.

It follows, from these and other considerations, that the factors governing suitability and actual exposure of fasteners in service cannot be compared. This does not invalidate test data for use in determining suitability, but shows that the data are mainly of service in estimating the margins to be allowed for actual work as compared to test conditions.

Pins

Holes in aircraft skins and substructures may cause conditions leading to fatigue, so that the current tendency is to use straight shanks for the pins because they give higher resistance to fatigue. Aluminium constructional work of aircraft type employs pins of titanium alloy beta III. This provides both high resistance to corrosion and low weight. The fastener head pressing flush with the structure has a collar driven in by upset on the opposite end. (See Fig 10.) It thus seals itself, the seal withstanding corrosion, stress corrosion cracking, and bad matching of countersunk holes

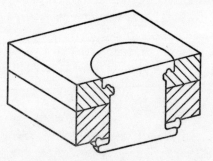

Fig 10 Stress pin fasteners: self-sealing characteristics of the head and collar of stress-pin fasteners inhibit stress-corrosion cracking and eliminate sealants

BEHAVIOUR OF PARTICULAR COMPONENTS

and heads. Sealing media are not required, and the claim is made that these pins enable an aircraft skin to be 10% thinner than usual owing to the higher fatigue strength they possess when so fastened. The fasteners are 22-60% lighter in weight than threaded and tapered fasteners. The degree of axial force transmitted to the pin shank within the hole causes it to expand and fill the hole tightly, so minimizing amplitude of stress.

Installation cost is 20-40% less than that of threaded fasteners owing to the simple form of the pins. Dimensional tolerances can be greater, maximum and stress amplitude being reduced.

Aircraft fasteners are always exposed to fatigue; but the stresses involved are not pure, which point is not always understood, though it certainly should be. Virtually all the fasteners used in aircraft engines are called upon to withstand stress rupture, and their ability to do this depends less on form and design than on the basic material of which they are made. Nevertheless, form and design cannot be ignored, since they are important in various highly preloaded applications. This means that all constructional joints needing to function at high temperatures must combine ample radii with gradual modifications of diameter, a high radius being adopted for the roots of threads, while wrenching surface must be correct.

Bolts

Large bolts are used in numerous applications, such as landing gears and attachments to airplanes of large size; bridges; rolling mills and similar static constructions. The severity of stress concentration in aircraft parts is much greater than in small bolts, and it is therefore far from easy to produce aircraft bolts that will meet the same specifications for fatigue as ordinary bolts. The small bolts have to withstand alternating stresses ranging from 4.5-5.0% of their minimum tensile strength as governed by the pitch diameter area. The minimum number of fatigue cycles to be withstood without failure is 45,000 and the average must be 65,000. Notched testpieces of large bolts are unlikely to conform to these requirements unless their rolled threads are formed *after* the metal has been heat treated. This is because high corrosion stresses then remain in the roots of the threads, and have to be overcome by the alternating tensile stresses before the root is subjected to tension. Hence, the tensile stresses encountered when a test-piece undergoes fatigue are held at a level adequate for an endurance limit double that of bolts thread-rolled *before* heat treatment. That is, the bolts thread-rolled *after* heat treatment have good fatigue strength, and the general conclusion may therefore be drawn that bolts of high strength for aircraft should also have compressive stresses embodied in their shanks during manufacture to protect them against static fatigue, stress corrosion, embrittlement by hydrogen and excessive tightening.

Load determining bolts are extremely useful, it is claimed, in determining by measurement the tensile load on themselves. Two types of mechanical bolts are used, the first, to show the relation of stress to strain in one and the same bolt under a comparatively high test load. The second shows the precise and uniform load on a joint that will eliminate uneven loading and so arrest too soon a constructional breakdown when the bolt is tightened to a pre-established load, or will give the joints longer service life after fastening. If the load-indicating bolts are regularly inspected they will reveal changes in preload of a specific fastener or form of fastener. Normally, the user's requirements are clearly indicated by drawings or specifications. More ample root radii in wrenching bolts lessen the effect of notches and so enable tension and tension fatigue stresses to be supported to a greater extent. The bolts, for instance, have to withstand cyclic loading ranging from high to low tension equivalent to 10% of the high tension load. The variations in fatigue

BEHAVIOUR OF PARTICULAR COMPONENTS

test load level factors are caused mainly by variations in the form of the threads.

Thread rolled bolts need to be competently rolled and as smooth on the surface as possible, as well as free from surface cracks, fissures and other defects. Given these requirements, such bolts can be favourably compared with the ordinary screw cut bolt. Rolling strengthens the root of the thread and gives the same type of residual compressive stresses as shot peening—the throwing of showers of iron and steel shot at high velocity by an air blast against the surface of a metal, to strengthen it against fatigue failure. It is a cold working process in which small overlapping indentations pre-stress a thin surface layer in compression. It is not the same as shot blasting.

The formation of a root with a broader and less abrupt radius enhances this effect, and these advantages obtain whatever the type of metal used for the bolts.

The fatigue stress encountered by bolts is principally in those designs embodying soft flat sheets such as gaskets or flanges. The hardness of the bolt materials, which are usually steel, give equal fatigue strength for equal hardness as long as identical surface conditions obtain. The steel used is less important than the condition of its surface. High carbon steels are seldom used for bolts because of their surface imperfections such as notches. A medium carbon steel is more suitable. The bolt fatigue limit can be increased to some extent by case carburizing or cyaniding, but these surface treatments are seldom employed, as the hard outer face they produce renders the steel less notch resistant.

Plating of the surfaces of bolts with chromium nickel, etc, lowers the fatigue strength except when the fasteners have to withstand only low stresses and the plating is purely ornamental. On the other hand, plating with zinc and cadmium does not greatly lower the fatigue strength, but is not often adopted because it may on occasion have a somewhat undesirable fatigue effect.

Bolt failure from fatigue occurs in most instances at the threaded area, or close to the leading thread within the nut. The concentration of stress leading to this result can be diminished by lessening the shank section just above the threaded area, or by exposing more thread between the bearing surface of bolt and nut. A good steel bolt should withstand up to 10^6 cycles before failure when subjected to a bending stress of 570-585 MN/m^2 (37-38 tons/in^2), the lower figure being for carbon steel, the higher for an alloy steel. Exceeding the maximum clamping stress is a com-

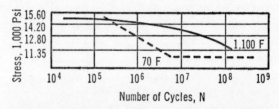

FIG 11 Effect of temperature on superalloy fatigue strength: bolt connections of A286, an ironbase superalloy, have more fatigue strength at 1,100°F than at room temperature. (1,000psi = 6.895 MN/m^2.)

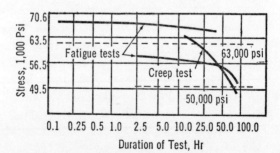

FIG 12 Fatigue range and creep tests of bolt connections indicate that resistance to both effects drops as time lengthens. Here, cyclic fatigue stresses were imposed on tensile specimens held under steady stresses of 50,000 and 63,000psi. (1,000psi = 6.895 MN/m^2.)

BEHAVIOUR OF PARTICULAR COMPONENTS

mon cause of looseness in bolts and leads to fatigue failure. (See Figs 11 and 12.)

Lock Nuts

Fasteners particularly requiring ample radii, high thread root radius and correct wrenching surfaces are self-locking nuts, jet engine tie-rod bolts, and load-rated ring-locked studs. Test procedures have been standardised for precision aircraft fasteners of this type and cover various means of producing a normal failure curve of fatigue stress against the number of cycles. There are no other existing criteria for determining the functional suitability of a specific fastener or form of fastener.

Lock nuts to carry loads, remain in position, and stay tight, have to meet conditions such as axial tensile stress, or wrenching and locking torques, and be trustworthy as regards permanent set durability, ability to be re-used, vibration, temperature and fatigue. They must above all withstand tension fatigue and the breakdown it may produce. The most effective means of preventing fatigue with these components is to keep the load on the threads constant. It has been demonstrated that the self-locking nut improves the fatigue life of mating bolts by as much as 25-50% endurance level. The bolt when tested wears before any nut impairment has taken place; and when the worn bolt is replaced, the performance of the assembly returns to its original level as represented by values of torque. The fatigue performance of the self-locking nut is also enhanced by proper choice of lubricant and by great attention to its surface finish.

CHAPTER 5

TESTING FOR FATIGUE - HIGH TEMPERATURE TESTS

TESTING FOR FATIGUE

The literature relating to fatigue is enormous in bulk and extent, but much of it is necessarily concerned primarily with theory and hypothesis. Research itself is intensive and cannot be summarized as a totality because results are not always conclusive, are occasionally ambiguous, and are not always accepted. Modern manufacturers and research workers use exclusively machines subjecting a component, sheet or plate, to 10^8 cycles of stress and these run for up to a month without pause. Fatigue failure over a wide range of loads and cycles is plotted until the fatigue properties of the metal under test are determined completely and with precision. The components are subjected to constant stress, or greatly varying degrees of stress, within carefully pre-selected limits. In many instances saline solutions are introduced to initiate corrosion, so that the corrosion properties of the metal can be established. In others, additional factors are linked with fatigue.

In the following notes, some of the more important research results are given in brief so that the reader may have some indication at least of the directions research is taking and why. This is particularly necessary for those employed in the aircraft industry.

For example, it is not now fully accepted that testpieces of ferrous metal, suitably finished by careful polishing, and having a Brinell hardness number below 500, have a bending fatigue limit at a considerable number of cycles amounting to about 50% of the tensile strength. This is said to suit the data for low and medium strength cycles reasonably well, but has no application to steels of high strength or to nonferrous metals. The fatigue strength in axial push-pull tests is found to be normally below that determined in rotating bending, which is itself below that in repeated bending. It is essential that in tests of this type the data obtained should be adjusted to take into account the effect of size.

Crack Growth and Loading Conditions

A variable function of modern fatigue research work is the ascertainment of the connection between the growth rate of a fatigue crack and the amount of stress, especially in metals for constructional work. By means of the data obtained it becomes feasible for the designer to adapt the information to a wide range of crack lengths, stresses and geometries, so determining the loading conditions for every specific design. In this way he is enabled to establish the proper stress expression identity and make up a body of facts covering small increases in crack growth. He can then draw a curve to show the connection between the crack length and the number of cycles that will cause failure at the operating stress of the component in question.

It is probable that the rate at which a fatigue crack develops depends on the degree of plasticity at the tip of the crack. Fatigue is strongly associated with hysteresis, which shows the degree of permanent set on the reversal of stresses.

Reference has earlier been made to constant life diagrams, which provide data on every type of stress from full reversal to differing tensile loads, and also give facts corresponding to the metal as it will be employed in service.

Scatter

The results given by tests often show considerable variations, usually termed 'scatter'. This is inherent in resistance to fatigue, and it has been firmly established that a number of metallurgical factors play a part in it. For example, it may arise from variations in the grain size of the metal, its composition or heat treatment. Moreover, there are differences between the various testpieces, while the fatigue strength of any one alloy made to a particular specification

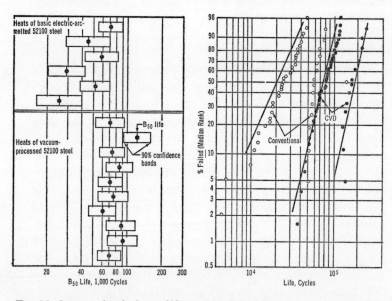

FIG 13 Scatter in fatigue life tests of hardened steel specimens representing different steel making processes *(left)* showing unknown factors in durability. It was then discovered *(right)* that individual specimens from all of the steels tended to fall into three separate groups—short, intermediate, and long-life—with the better performing steels exhibiting fewer of the short-life-type specimens

TESTING FOR FATIGUE

may greatly differ by reason of varying working practices. As far as the writer knows, the only facts confirming this latter effect relate to aluminium alloys produced by extrusion, but the data obtained have no applicability to steels or other nonferrous metals.

The basic facts of fatigue fracture throw little light on actual fatigue problems, but there is an enormous mass of theoretical matter which engineers and metallurgists have to grapple with as best they can.

Taking the problem of scatter more generally, it is recognised that it depends to a considerable extent on the finish of the piece undergoing test. For example, it is greater when the surfaces are smooth and polished than on those that are notched. It is greater on high strength metals than on those lacking in strength. It increases as the stress range increases, and may be extremely marked. Scatter in the stress range for a specific endurance may be much lower in degree than at a specific range of stress. (See Fig 13.)

It is frequently asserted that if fatigue tests are to prove a useful factor in ascertaining the performance of metals beforehand, it can be only by the efficient use of statistics and analysis by statistical means. This alone will enable forecasts to be quantitative and the trustworthiness of metals to be foreseen. These methods also provide a fairly good assessment of results when the data given by the tests are considerably at variance. This is useful when determining, for example, the influence of heat treatment or production processes on fatigue life.

Because of scatter it is held that a large number of tests must be carried out for any worthwhile data to be obtained and statistically considered. Eight is the suggested minimum. S-N curves are normally taken over 10^5-10^8 cycles. These are curves drawn to indicate the relation of stress to the number of cycles that can be withstood before fracture takes place. (See Fig 7.) They are obtained by plotting the stresses on logarithmic or linear co-ordinates, and the number of cycles on logarithmic co-ordinates. The method of plotting chosen should suit the data required. Some use of these curves is made to ascertain the anticipated service life of components under stresses greater than the previously indicated fatigue limit, and in this instance the forecasts made will have greater value. The results are plotted as hours before failure, or miles of use, so as to show the rates at which failure will take place during the design life of the machine.

A degree of plastic deformation frequently takes place, the precise figure being governed mainly by the stress. It is advisable to provide some method of cooling those metals revealing high hysteresis where dispersal of the heat produces a considerable increase in the temperature of the testpiece. Consequently, the testpiece temperature should be measured by a suitable pyrometer.

Progress of Techniques of Fatigue Research

The advance of knowledge of metal fatigue has arisen from the development of new and improved scientific instruments such as the electron microscope, x-ray diffraction, thin film techniques, and those of advanced metallography. New and sensitive testing equipment, new microscopes, the highly skilled interpretation of metal fractures aided by colour photography, new etching procedures, have all played their part. Laboratory testing machines have continually diversified, and now range from a single fixture for a specific piece of work to intricate machines of large dimensions having an extensive range of capacities. Variable loading fatigue tests are often advisable and are used for complete machines or major assemblies. A system of sophisticated tape records actual field parameters and plays them back through a control system as a means of producing variations of loads.

It is considered specially important that every item of a fatigue test shall be exactly typical of the component or

TESTING FOR FATIGUE

piece needing to be tested. Every detail of construction has to be introduced into the testpiece where it correctly occurs. End attachments are extremely critical, and the load must be applied to the testpiece at a lower concentration of stress than is found at any other point of the testpiece itself. The placing of the load axis right through the testpiece, and particularly at its end connection, is essential so that undesirable bending stress does not cause too early a fracture. Both the testpiece and the final S-N testing should promote the precision with which the metal is evaluated and also the correctness of later decisions.

Constructional breakdowns are often caused by 'fretting', the wear that takes place when two metals rub together with a reciprocating motion of limited amplitude. The data given by variable and constant amplitude loads is affected by different factors, so that the connection between constant amplitude and variable amplitude tests is much more intricate than supposed.

TYPES OF TESTING MACHINE

The five principal types of testing machines are the rotating-bending mechanical machines, repeated-bending mechanical machines, axial hydraulic closed-loop machines (axial loading direct stress), repeated-bending electromagnetic resonance machines, torsional-mechanical resonance machines, and axial-thermal fatigue machines. Space does not allow of more than a brief description of the principles involved in these six classes.

Rotating-Bending Mechanical Machines

Rotating-bending mechanical machines employ a cantilevered fatigue testpiece which has to withstand a bending load of constant magnitude, and is rotated to give intricate reversal of bending stress from maximum tension to maximum compression for each test.

Repeated-Bending Mechanical Machine

A crank mechanism is embodied in this to provide a constant bending deflection. Maximum strain range is applied to only the top and bottom surfaces.

Axial Hydraulic Closed-Loop Machines

These make use of an electro-hydraulic control system and are extremely versatile. The hydraulic components can be sized for any testpiece ranging from 6.3mm (0.25in) diameter up to large welded constructions. The machine is normally regulated by a strain gauge located on the component undergoing test. This gauge indicates and so controls the servo valve governing the oil flow to the power cylinder. The machine can run to either constant load or constant displacement, and the regulating indications may be taped signals field recorded duty cycles.

Repeated-Bending Electromagnetic Resonance Machines

In these an electromagnetic system develops the cyclic force, the mechanical natural frequency of the tuning fork being close to or equalling the frequency of the electromagnet. By electronic controls, a constant amplitude is maintained and monitored by a strain signal from the crankshaft.

Torsional-Mechanical Resonance Machine

Here a rotating out of balance weight develops a centrifugal force which induces a torsional moment in the testpiece. The system normally operates at resonance with the electric amplitude controls. It is claimed that resonant operation is advantageous in that large forces and moments are set up by small exciting forces.

Axial-Thermal Fatigue Machine

This is used for thermal fatigue testing and employs mostly a plain testpiece set between the top and bottom plates

TESTING FOR FATIGUE

of the fatigue machine. The thermal expansion and contraction of the restrained testpiece create the necessary forces. Normally, heating is by electrical resistance, and cooling by still or moving air. The testpiece is almost wholly restrained.

This form of test is specially suitable for the metals of diesel engine cylinders, the amount of stress being modified by altering the range of testing temperatures. From the test it is possible to obtain facts concerning the influence of the range of cycles up to failure. The test must be valid for either the single part of a machine or construction, or for the entire assembly or construction.

A variable loading fatigue test may be advantageous for complete machines or large assemblies. It is necessary that a definition of fatigue failure be provided for each series of tests.

Ambiguity

It should be noted that ambiguity is sometimes found in tests carried out on rotating beam machines, which tests can be performed only with fully reversed loads. An axial stress testpiece is superior to the conventional since the whole of the cross section is at stated stress. Nominal stress is determined by simple theory, which does not take into account modifications of stress conditions set up by holes, grooves, fillets, etc. Each testpiece is subjected to a specific fluctuating or alternate loading between the number of cycles producing fracture (N).

Important Tests

Increasing weight is being given today to low cycle fatigue and thermal stress fatigue, while, as indicated earlier, the effect of a liquid metal environment and of nuclear radiation on the fatigue strength of metals employed in various fields is also being persistently studied. The effects of fatigue and creep combined in metals for power station plant and for the blades of gas turbines are also regarded as of great importance. The detailed design of joints under fatigue loading, the failures of bolted, riveted and welded steel and aluminium, the problems of fatigue in supersonic flight, satellites and guided missiles, and in new chemical processes are all being investigated.

Ultrasonic Tests

These are being used to determine the fatigue response associated with nonmetallic inclusions of differing magnitude, as in high strength steels (see Fig 14). The method

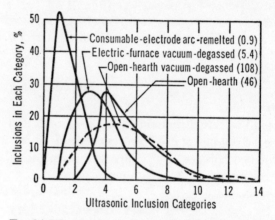

FIG 14 Ultrasonic tests and cleanliness: tests revealed that AISI 4340 made by different methods varied appreciably in cleanliness. Numbers in parenthesis give comparative cleanliness ratings

adopted is to produce uni-axial fatigue testpieces of different degrees of freedom from these inclusions as measured ultrasonically. The testpieces are heat treated to about 2,500 MN/m^2 (160 tons/in^2) tensile strength. In the actual testing operation transverse and longitudinal testpieces are loaded in tension-tension and tension-compression by a universal fatigue tester, the maximum stresses being about

TESTING FOR FATIGUE

20% over the fatigue limit to give adequate plastic strain and developed breakdown inside a specific length of time.

The fewer the inclusions, the longer the test life of the piece; the improvement produced by greater freedom from inclusions being about twenty times. A higher rate of failure under tensile stress appears to occur with increase in the size of the inclusions. The fatigue resistance of a plastically strained metal can be gathered in advance from the parameter of fracture ductility, itself correlated with the absence of inclusions in a metal of specified hardness and microstructure.

According to Cellitti and Carter, this relation is expressible as $\log y = A \log x + B$, hence $\log y = 16.32\ x^{-0174}$ where y is the index of cleanliness rating and x the number of cycles leading to failure. This relation is not valid at extremely low lives (below 10^4 cycles), a more realistic limit lying between that value and 10^6 cycles, a higher precision being attainable as the cleanliness rating for inclusions declines. In both the above mentioned instances the material was a 4340 carbon steel.

HIGH TEMPERATURE TESTS

The fatigue behaviour of stainless steels as affected by time and low cycle has been examined by Berling and Slot. In brief, they discovered that fatigue life in relation to the sustained number of loading cycles generally, but not in terms of the elapsed time, was lessened when temperatures were increased from 430-650°C (800-1,200°F) and up to 816°C (1,500°F) as well as by a rise in rate of strain from about $4 \times 10^{-3} s^{-1}$ to strain rates lower by factors of 10-100. There is reason to believe, however, that lower temperatures than these may also have a bad effect on strain rate, but the tests concerned were not carried out below 650°C. In these tests great importance was attached to precise control of testing conditions, and the use of push-pull loading of solid round testpieces so as to give a uniform distribution of strain in the zone of fracture. Local control of temperature and strain was achieved by using an hourglass gauge section, while constant strain rate during the loading cycle was maintained. On the basis of preliminary results, it was concluded that the rate of strain reversal between holding periods must be taken into consideration as a test parameter, so that a well defined cyclic loading mode is an important objective in holding time conditions.

Columbium (Nobium) Alloy Test

The low cycle fatigue properties of heat resisting alloys in vacuum at high temperatures have been studied by Swindeman, taking an alloy known as D43. The testing temperatures ranged from 20 to 1,204°C (70-2,190°F), with cyclic plastic strains from 0.3-0.4%. The plastic strain resistance increased as temperature rose to show a maximum about 871°C (1,599°F). Up to 1,024°C (1,870°F) the cyclic strain resistance was and continued to be superior or equivalent to that at room temperature. After the tests had been completed, the testpiece showed ductile fractures across the grains at all temperatures. The test was carried out on a specially designed machine operating a push-pull load and with a controlled stroke. It would operate at temperatures up to 1,400°C (2,550°F), and would take loads up to 4.45×10^4N and frequencies from static to 10Hz. The test section was 12.7mm long and 5.18mm diameter.

High Purity Nickel Test

Stegman and Shahinian have studied the influence of temperature on the fatigue of nickel at varying oxygen pressure, under conditions of reversed bending and at temperatures of 20, 300, 425 and 550°C (70, 570, 790 and 1,020°F). At each of these temperatures it was found that a transition zone exists in which there is a comparatively abrupt decline in fatigue as oxygen pressure increases, until a high level of pressure termed the maximum transition pressure is reached. This pressure is roughly the same at 300°C as at 20°C, but increases progressively as temperature increases still further. At high temperatures, the rise in transition pressure is considerably greater than would be predicted. This is regarded as caused by oxidation along the surfaces of cracks, lessening the number of molecules able to reach the advancing tip of the crack. A higher strain reduces the degree of environmental effect at lower temperatures of 300 or 550°C without any visible influence of strain on the maximum transition pressure. The effect is believed to be connected with strain hardening at the lower temperature not existing at the higher temperature.

Test of Creep and Low-Cycle Fatigue at High Temperature

Wells and Sullivan have been studying the interaction of low cycle fatigue with a period of creep in each cycle for a wrought nickel base superalloy known as Udimet 700, used for gas turbine blades and discs. It was found that this interaction is predictable as regards linear cumulative damage if it is assumed that in static creep a large crack initiation life exists. The maintenance of compressive strain appears to be more injurious than that of tensile strain, while compressive creep pre-strain is more harmful than an identical strain in tension.

The rate of creep is lower in compression than in tension and declines with strain. Deformation suggests heterogeneity and planar effects, accompanied by large concentrations of stress where dislocations accumulate against boundary particles. Some points, such as grain boundary cavity growth and creep coalescence and fatigue, as well as the relevance of the static creep test to creep fatigue interaction, are still in doubt. However, a mechanism has been suggested for the interaction of creep and fatigue damage.

Creep-Rupture-Ductility Test

The influence of creep rupture ductility and holding

time on the 540°C (1,000°F) strain fatigue behaviour of a 1% chromium, 1% molybdenum, 0.25% vanadium alloy steel has been studied by Krempl and Walker. Steel of this type was heat treated to low and high levels of creep rupture ductility using completely reversed uni-axial strain fatigue tests, the holding times being 1, 10 or 60 minutes at the maximum tensile strain point of the cycle. The total strain range was either 0.5 or 1.0%.

Cycles to failure declined notably with increase in holding time, the fall being greater for the 0.5 than for the 1.0% strain range. The holding time effect was much more marked in the lower-creep-rupture ductility metal. The character of the cracks moved from across the grains to between the grains as the maintenance time increased, showing that creep rupture damage took place during the maintenance periods. This movement was more complete for 0.5 than for 1.0% strain range tests and its extent was greater in the low than in the high creep-rupture-ductility steel.

Both steels softened under strain during cycling, there being an uninterrupted fall in the total stress range. The greater strain softening was found in the high ductility metal. The inter-relation between the properties of the metal was considerable in the tests and it is suggested that relaxation, creep and cyclic strain softening (hardening) are inter-related. No unique law has yet been found to represent the observed behaviour.

Effect of Temperature on Creep

The effect of this in tubes of pure aluminium has been the subject of research by Kitagaw, Jaske and Morrow. It has been shown that creep accelerates with repeated reversals of stress at high temperatures, while deceleration occurs at low temperatures. The possible decline in fracture life caused by the acceleration may result from reversed stressing at high temperature, and has been studied in relation to creep rupture, high temperature fatigue and thermal

TESTING FOR FATIGUE

fatigue. A temperature of about $0.4\ T_m$ is believed to be that above which creep deformation under reversals of stress may injure the fracture properties when the metals are subjected to repeated stress reversals.

High-Temperature Low-Cycle Fatigue Tests

According to Carden and Slade, the thermal fatigue behaviour of an alloy for heat resistance, known as Hastelloy X, in thin-walled tube bundles can be ascertained from controlled strain tests applied to the thermal cyclic behaviour when the tubes are joined to a rigid plate. The conclusions tentatively reached are that rise in temperature, both isothermal and cyclic thermal, reduce the strain-range fatigue relation. There is more likelihood of buckling for the negative type loop tests. Over the range of 36 to 420 cycles/hr, no marked frequency effect was observed. The thermal fatigue behaviour of the tube bundles is related to low-cycle high-temperature fatigue behaviour of the metal. Progressive instability in thermal cycling of the bundles allows local strain concentrations to develop. The ductility of the alloy in relation to temperature is not low enough at the minimum to produce a serious impairment of the high temperature fatigue behaviour.

Low-Cycle Fatigue Life of Steels for Pressure Vessels

De Paul and Pense have tested the low cycle fatigue behaviour of steels for pressure vessels, particularly in relation to the influence of microstructure and alloying elements. In these vessels fatigue is a primary factor. They show an average service life of up to 10^5 cycles, so that these influences are of great importance to both makers of the materials and suppliers. The strains experienced within the range of 5,000 to 10^5 cycles are different from those usually taken into account in fatigue work on endurance limits, being greater and usually resulting in a degree of plastic deformation of the material. The object of this series of tests was to determine the effect of microstructures given by practical heat treatments on the room temperature fatigue resistance of pressure vessel steels.

The conclusions reached suggest that normalizing gives carbon and low alloy steels a better low cycle fatigue resistance than quenching and tempering to the same tensile strength over the entire cyclic range. High strength steels are better at lower strain levels, but do not inevitably give the best fatigue resistance over the whole range. A higher alloy content gives greater fatigue resistance above 20,000 cycles, but carbon steel unalloyed is better in the 5,000 cycle range.

If microstructure is constant, the quenched and tempered microstructure, giving greater tensile strength by variation of tempering treatment, moves the low cycle fatigue resistance to higher permissible strains for any specific life.

Fatigue Properties of Titanium Alloy

A commonly used and popular titanium alloy containing 6% aluminium and 4% vanadium, has high tensile, creep, notch stress rupture and impact properties at high temperatures. Not much was known, however, about the effect of different microstructures on its fatigue properties. Accordingly Bartlo investigated, coming to the conclusion that microstructural variations changed the tensile and fatigue properties. The most satisfactory combination of tensile and fatigue properties was obtained when the alloy had a microstructure of fine-grained alpha beta type or microstructures in which fine primary alpha and martensitic alpha were mixed. Heat treatments coarsened grain structure, and reduced fatigue strength as well as endurance ratio. Martensitic alpha as the principal phase in a microstructure, developed by beta quenching, showed good fatigue strength, but had low tensile ductility.

The endurance ratios of all smooth testpieces were from 0.40 to 0.62 with about 60% of the treatments having endurance ratios of between 0.58 and 0.62. Air cooling or furnace cooling after beta treatments gave endurance ratios on the low side of this range, whereas beta quenching and most of the alpha beta treatments gave fatigue strength and endurance ratios between 0.5 and 0.62.

The fatigue strength of the notched testpieces was about 50% of the unnotched fatigue strength. No marked changes were seen when notched fatigue data were compared with the microstructures.

CHAPTER 6

MODERN USA AND USSR RESEARCH INTO FATIGUE FAILURE

To compress into a small book such as this the entire mass of information regarding tests connected with problems of fatigue failure by plastic deformation would be impracticable. In England, the United States, China, Europe, the USSR, work is steadily going on. Among the more interesting researches of recent years, a few have been chosen from abroad, such as the use of polarized light in fracture mechanics phenomena. An accelerated method for determining the fatigue limit, the effect of stress concentrations on fatigue strength, are among the particular items summarised here.

Cross Polarized Light Technique

Noritake, Wals and Roberts in America have introduced a technique for using cross polarized light to delineate fracture zones clearly for photography. In this, white light travels from the source through a polarizer, known as the first, to a glass plate which deflects it 90% on to the testpiece. The light beams bounce back from the surface and thence through the glass plate to a second polarizer and a camera lens to strike the film. (See Fig 15.) The crossing of

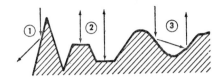

Fig 15 Effect of polarised light: light striking the surface of a sample is (1) deflected to the side, (2) reflected 180° directly, or (3) reflected 180° indirectly

the polarized filters prevents extraneous light rays from striking the film and masking important details. This crossing is achieved by placing the second polarizer vertically so that its polarization plane is parallel to the horizontal edges of the rectangular glass plate. The second polarizer is set horizontally with its polarization plane at 90° to that of the first filter. The glass reflector-refractor is at about 45° to reflect the maximum amount of polarized light possible on to the testpiece surface. This cross polarized light shows up differences among fractured surfaces, and zones can have their dimensions determined as well as their distance from the notch, which facilitates calculations for fracture mechanics.

A technique using single polarized light has also been developed to show up cleavage facets of coarsegrained fractures. More light is reflected from the fracture zones than in the crossed polars technique, and gives an unusual highlighting effect.

Mechanism of Crack Formation

Oding in the USSR, has suggested two possible mechanisms for the formation and growth of fatigue cracks in metals. One is a diffusion mechanism in which the dominant role is played by vacancies in the crystal, and the second is a diffusionless mechanism in which the interaction between dislocations predominates. Fatigue failure may be caused by either, according to the cyclic stressing conditions such as temperature. Finally, the character of the metal also determines which of the two mechanisms is at work.

Structure Energy Theory

Many previous theories of metal fatigue were founded on structural factors only, or on energy considerations alone. Consequently it was difficult to make practical application of them. Ivanova has put forward a somewhat complex structure-energy theory which suggests that both the modifications of microstructure occurring in the metal, and the values of the energies required for the processes during the stages of fatigue, should be taken into account. The theory is founded on the notion that the energy of failure is not dependent on the manner in which the energy is applied. Consequently it becomes possible to compare the failure energy and the latent heat of fusion in plastic deformation leading to fatigue.

Fatigue Crack Development

Investigations into the growth of fatigue cracks are said to show that the development of such a crack takes longer as the testpiece size increases, as the stress concentration increases, and as the strength of the metal decreases. The rapidity with which a fatigue crack propagates, declines as the testpiece size declines and the limiting stress decreases.

Fatigue and Hardened Steels

Much progress is recorded in the fatigue testing of hardened steels. Some evidence has been produced after experiments on ductile metals, but work on fatigue of these metals, and especially of hardened steels, is by no means exhaustive. The problem is significant, however, because numerous steels of high strength are required in mechanical engineering, and could be used much more widely if their cyclic stress behaviour were more clearly known. Some recent research has shown that the fatigue limit for these steels is now known, and indicates a continuous relation between fatigue strength and load application time within the range 500-10^9 cycles, expressible as a straight line in—log N coordinates. It seems that hardened steel cannot be tested to this specific number of cycles. In general, fatigue failure with the steels takes place somewhat differently from that with ductile metals.

Fibre Direction in Steel Parts

Fatigue failure usually takes place in a brittle manner, and is believed to be associated with shear stresses. Failure caused by cyclic contact stresses is characteristic of balls for ball bearings, rollers and cages, gears, rails, wheels, steel tyres, the journals of cold rolling mills, and many other parts. Cyclic loads producing variable stresses result in dislocations causing numerous vacancies. These collect into clusters and form submicroscopic pores and possible cracks. This chiefly results in hardening the surface layers during contact loading.

Investigations into a large number of different fatigue failures in the type of components mentioned caused by cyclic bending, twisting, tensile and compressive stresses, have revealed, it is claimed, that the fibres of the metal in those cross sections most severely stressed must lie parallel to the maximum tensile stresses or strains, but must not follow the maximum shear stresses. Transverse fibres in axles and shafts, for example, reduce fatigue strength in torsion as compared to longitudinal orientation, which gives the best results.

Accelerated Method of Determining Fatigue Limit

Experiments have indicated that fatigue limit can be calculated with an accuracy adequate for practical purposes from the critical fatigue stress. The method proposed to achieve this greatly reduces the time necessary to establish the fatigue limit, since only a limited number of testpieces need be tested to failure. The main feature of these experiments is the establishment of new fatigue criteria, such as the critical number of cycles, or of cycles leading to failure at the critical stress, and the cyclic fatigue constant.

Additional work shows that this method can be used to determine the fatigue strength for solving various technological problems with at least promising results. It is, however, considered that the data obtained should be regarded as provisional only, practical application being restricted by the lack of precise fatigue properties for most alloys, and also of definite information on the limitations of the method, such as size factor, experimental conditions, etc.

New Vibratory Testing Machine

A new method of carrying out vibration fatigue tests has been introduced in the USSR. Its basis is that these tests should be performed with greater damping, set up artificially by means of an external loading, at some point in the mechanical system apart from the testpiece. The loading must be proportional to the displacement velocity. Mechanical systems of most vibratory machines have only small damping, so that a high sensitivity at resonance to variations of the forced frequency is inevitable. Stabilization techniques founded on obtaining a stable forced frequency are intricate and not always trustworthy. The load therefore varies, a good deal of scatter is seen in the results of vibration fatigue tests and the accuracy is not great.

In the new method, damping is attained by the application of an external mechanical concentrated load proportional to the linear displacement velocity. Because the damping is high, less sensitivity to frequency variations exists, so that special excitation sources become necessary. Thus, in electrical machines of resonant pattern normal mains supply can be used. The new method gives the essential stability of stress conditions, afterwards increasing the accuracy of fatigue property determination.

Notch Sensitivity of High Strength Steels

On the basis of experiments, it is suggested that strength and hardness are not the primary causes of notch sensitivity in steels, and that there is no direct relation between the two. Experiments are said to show that workhardening does not affect the general relation between notch sensitivity and strength, but influences only the absolute value of the sensitivity index. Studies of the hetergeneous structure of the high strength steels will be necessary before additional explanations of increased damping strength are obtained.

Size Effect and Corrosion Fatigue

Hitherto, studies of this factor have been restricted to structural steels and the effect of tap water upon them. The data obtained were contradictory. In consequence, investigation was undertaken into the influence of a more corrosive medium, sea water, on carbon and alloy constructional steels, austenitic stainless steels, and two kinds of brass.

The data resulting confirmed that corrosion fatigue failure is determined by the interaction of corrosion and mechanical factors. The conditions being equal, the effect of these on material, frequency of cycles, form and dimensions of the testpieces, and the corrosive medium, was found to be governed by the number of cycles on which the tests were founded.

Damping Capacity and Internal Friction

Over the past twenty years many researches have been carried out into the degree to which the internal friction of pure metals and alloys depends on their temperature. Much has been done to explain how a metal behaves under load at varying temperatures; while at the same time internal friction has itself been better explained. It became feasible for a relation to be established between the strength of a metal and its internal friction, and to this end tests were carried out; while also the feasibility of studying work-hardening, and later decline in strength, by the internal friction method was also taken into account.

It was found that the internal friction at high temperatures can be employed as a basic property of the strength of a metal in the temperature range 1,000-1,100°C (1,830-2,010°F). Also within this temperature range internal friction varies with the mechanical properties governing the strength; but at high temperatures the sensitivity of this type of friction to microstructural changes occurring in the alloy under stress is much greater that that of other microstructure-sensitive properties.

It is suggested, therefore, that the internal friction method is safe for laboratories studying the mechanical properties of metals, and the method is already in use. On the other hand, internal friction is a complicated phenomenon governed by its high sensitivity to microstructural change in the metal, etc, and it is recommended that an internal friction theory should be developed.

External Medium and Fatigue Strength

Time is a major factor in relation to the effect of external media in static or fatigue tests, so that much time is needed for the various changes in the mechanical properties of metals to reveal themselves. Corrosive media almost always affect the long term and fatigue strength, and the effect on fatigue strength is greater as the thermodynamic instability of the metal increases. Tests with media that combine purely mechanical effects with surface and diffusion phenomena have been carried out. Adsorption occurs for most media and precedes corrosion and diffusion effects, so that it is a universal factor, it is claimed, leading to reduced surface energy, with deformation and failure more rapidly effected, and even spontaneous disintegration at very small stresses, or even without external loading.

Another universal factor is said to be defective microstructure, including submicroscopic to microscopic defects, dislocations and vacancies predetermining the deformation of solids, and the concentration of defects close to the surface. Research has revealed many other valuable facts, and stresses the effect of machining on the fatigue of metals in active corrosive media, and the enhanced effect on the fatigue strength of metals of the technological processes preceding the final treatment. Surface rolling is suggested as a final treatment suppressing all changes arising from preceding treatments, and at the same time giving a uniformly workhardened surface, particularly for parts working in active media. This operation also increases the fatigue strength.

Machine for Studying Corrosion Fatigue of Metals

While developing new equipment for the study of corrosion fatigue on metals, the Railway Transport Institute of the USSR has enabled a machine to be designed and constructed for the study of fatigue crack initiation in testpieces. The machine comprises a mechanical arrangement, an electronic generator and a measuring system allowing periodic measurements of the full period of oscillation of the testpiece. The machine functions at resonance, the frequency being governed by the rigidity of the specimens and the applied weights. The testpiece is immersed in a corrosive solution so that the corrosion fatigue strength may be ascertained over a frequency range. The mechanism of the machine ensures that the stresses are constant over the gauge length of the testpiece. The aim of the tests is to obtain a relation between the oscillation period of the testpiece (all other parameters being constant) and the time.

GLOSSARY

BASIC FATIGUE TERMS

A = Stress Ratio.

Alternating Stresses. Stresses varying rapidly between maximum and minimum induced in a metal by a force acting alternately in opposite directions.

Basic Tensile Cycle. A pulsating cycle whose applied stress is invariably a tensile minimum ($=0$) and mean stress $=\frac{1}{2}$ max. (This term is not specially mentioned in BSS 3518, Pt 1 of 1962 governing fatigue nomenclature).

C = Cycle Ratio.

Cycle Ratio. The relation of the number of stress cycles of indicated character to the hypothetical fatigue life derived from the S-N diagram for stress cycles of identical type. It is usually written C.

Endurance. The number of stress repetitions varying in magnitude but not in sign.

Fatigue Terms. Detailed investigation of the causes of fatigue and its consequences did not begin in earnest until after 1946, when high frequency oscillations were found to be induced by the noise from large gas turbine and rocket engines, while missiles often failed owing to these high frequency vibrations. The following technical terms commonly used in the literature on the subject are given for the benefit of the reader coming to it for the first time.

Fatigue Life. The cycles of stress supportable in given circumstances before a metal fractures or breaks down. It is usually denoted by the letter N.

Fatigue Limit. That stress just too high for a metal to support when subjected to stress cycles of infinite number. Whenever the stress is not fully reversed, the mean or mini-

GLOSSARY

mum stress value or the stress ratio is given. This ratio is commonly expressed as S_a.

Fatigue Notch Factor. The relation of the fatigue strength of an unnotched testpiece having no stress concentration to the fatigue strength at an identical number of cycles having concentrations of stress at identical conditions. Usually written as K_f.

Fatigue Notch Sensitivity. A degree of agreement between K_f and K_t for a specified dimension and metal of a testpiece having a specified stress concentration size and form.

Fatigue Range. That range of stress not causing failure of a metal when acted upon by a cause or causes of fatigue.

Fatigue Ratio. The relation the fatigue limit bears to the tensile or static strength of the metal for cycles of reversed flexural stress. This value is usually determined by a special type of test.

Fatigue Strength. The greatest stress a metal will support without fracture for a specific number of stress cycles, the stress being wholly reversed within each cycle unless otherwise indicated. The stress may be equal in tension or compression and successive, or involve alternate tension and compression with intermittent periods of rest. The fatigue strength for a hypothetical stress value is usually denoted by the type of diagram known as an S-N curve (see Fig 7, which represents the precise number [N] of cycles producing failure.)

Fatigue Strength Reduction Factor. The relation the fatigue strength of a component or testpiece without stress concentration bears to the fatigue strength when such concentration is present. It is often denoted by K_t and has no meaning unless the geometry, dimension and metal of the component or testpiece are indicated, together with the range of strength. See Theoretical Stress Concentration Factor.

Fluctuating Cycle. One that remains of the same sign, but varies in magnitude.

Fluctuating Stresses. Stresses varying in magnitude but not in sign.

Fracture Toughness Value. The relation between the energy absorbed in the impact test and the toughness of a metal component in service. This is a generalisation not founded on scientific laws, and is essentially qualitative, but it is being increasingly used to indicate the critical intensification of stress at the tip of a crack that will lead to swift growth of the crack in a metal. Regarded as a simple property, equivalent to the use of tensile strength and yield strength, it is a constant to which the letters K_c are given.

K_c = Fracture Toughness Value.

K_f = Fatigue Notch Fracture.

K_t = Theoretical Stress Concentration Factor or Fatigue Strength Reduction Factor.

Limiting Range of Stress. The maximum stress range (mean stress = 0) a metal can undergo for an indefinite number of cycles without failure. As the stress range increases, the number that can be withstood declines. An alternative term is *Endurance Range*, which indicates twice the fatigue or endurance limit.

Loading Cycle is denoted by the formula $S_m \pm S_a$.

Maximum Stress. That amount of tensile stress a metal will bear when loaded under specified conditions until it fractures in a tensile testing machine. It is calculated from the tensile stress and the initial cross sectional area of the testpiece. In the tensile test this stress is expressed as positive (+). The term is usually denoted by S_{max}. It is sometimes better known as Ultimate Tensile Strength, and is calculated by dividing the stress by the strain in any elastic deformation.

Mean Stress. A point halfway in the stress range. If the stress is zero, the upper and lower limits of the range are of equal value, but one is in tension (marked +) and the other in compression (marked −). It is usually written as S_m, and when both limits are equal, can be calculated by adding the

GLOSSARY

maximum to the minimum stress and dividing by 2, that is $\frac{S_{max} + S_{min}}{2} = S_m$

N = Fatigue Life.

Pulsating Stresses. Those varying in magnitude but not in direction.

Q = Fatigue Notch Sensitivity.

R = Stress Ratio.

Range of Stress. The difference between maximum and minimum stresses in a single cycle. Usually written S_r, it equals the subtraction of the minimum from the maximum stress, the result being divided by 2, that is $\frac{S_{max} - S_{min}}{2}$.

Reversed Asymmetrical Cycle. When S_m is not 0.

Reversed Symmetrical Cycle. When $S_m = 0$.

S = Nominal Stress.

S_a = Alternating Stress.

S_d = Fatigue Limit.

S_d/S_u or S_n/S_u = Ratio of Fatigue Limit to Ultimate Tensile Stress (S_u).

S_m = Mean Stress.

$S_m + S_a$ = Applied Stress Conditions.

S_n = Fatigue Strength.

S-N Curves. Curves drawn to indicate the relation of stress to the number of cycles that can be withstood before fracture takes place. See p55.

S_r = Range of Stress.

$\frac{1}{2}S_r$ = Stress Amplitude.

S_u = Ultimate Tensile Stress.

Stress Amplitude. One half of the stress range, usually written as $\frac{1}{2}S_r$.

Stress Cycle. A series of repeated stresses in which the state of a metal at the end is the same as at the beginning. Generally the cycle is a recurring one, numerically defined as $M + R$ where M is the average stress and R the stress or fatigue range.

Stress Modulus. Stress divided by the strain in any elastic deformation.

Stress Ratio. The modulus of two specific stress values in a stress cycle. It may be written A or R, and represents the minimum stress divided by the maximum stress. Two frequently employed ratios are $A \text{ (or R)} = \dfrac{S_{min}}{S_{max}}$ or $A \text{ (or R)} = \dfrac{S_a}{S_m}$ which is the ratio of stress amplitude to mean stress.

Theoretical Stress Concentration Factor. The relation of the maximum stress in the stress concentration zone produced by a notch (determined by the theory of elasticity, or by experiment) to the corresponding nominal stress. See Fatigue Strength Reduction Factor.

$+$ = Tensile Stress.
$-$ = Compressive Stress.

METALLURGICAL AND OTHER TERMS

Ageing. The structural change taking place in certain metals at atmospheric temperature or more rapidly at higher temperatures. Certain values such as proof stress, maximum stress and hardness are increased, but ductility is lessened.

Annealing. Increasing the temperature of steel to a particular point, maintaining it at that temperature for about 1hr/in of cross section, then cooling slowly *in the furnace.* The object is to soften the steel for easier machining and working.

Austenitic. Of steels composed of austenite, an allotropic form of iron, which remains stable at normal temperatures.

Blowholes. Round or elongated smooth-sided gas-containing cavities in solid metals, particularly ingots and castings.

Brittle Fracture. The failure of a metal by cleavage owing to the exceeding of cohesion.

Cleavage Planes. Planes along which fracture is easy, owing to the elongation of the grains of a metal subjected to a stress parallel to the faces of the system to which the grain or crystal belongs.

GLOSSARY

Crackless Plasticity. The submicroscopic action of giving way at the stress zone without the formation of a crack. An alternative term is *Damping*.

Creep. Gradual but continuous deformation of a metal when under steady load.

Critical Point. The point or temperature indicating a change of phase in steel. It is signified by a sudden liberation of heat during cooling or absorption of heat on heating.

Crystalline Fracture. A bright and glittering fracture occurring in metals when the fracture runs along the cleavage planes of the crystals.

Cyaniding. Introducing carbon and nitrogen into steel at the surface by heating it in a molten bath of cyanide or cyanide salts. The treatment gives the steel a thin outer case of considerable hardness.

Damping. In vibrating mechanical systems, the resistance of the motion by influences taking up energy, limiting and possibly stopping the vibration.

Decarburization. Heating ferrous metals in a medium that reacts with their carbon content, so robbing them of some of their carbon and forming 'soft skin'.

Dislocation Density. The state of density of metal crystals allowing vacant spaces to accommodate spare atoms under stress. This is a theoretical explanation of plastic deformation, creep, diffusion and other phenomena in metals.

Elastic Hysteresis. See Damping.

Ferritic Steels. Those not forming austenite when heated.

Fretting. The wear on metals rubbed together with a reciprocating (to and fro) motion of limited amplitude.

Hypereutectoid Steel. Steel with carbon content of about 0.83%.

Hypoeutectoid Steel. Steel with less than 0.83% carbon.

Hysteresis. The difference between the heating and cooling critical points caused by the lagging behind of physical changes at the temperatures of heating and cooling.

Internal Friction. The ability of a solid metal to disperse applied mechanical energy as heat.

Laps. Defects on the surface of mechanically worked metals caused by part of the metal folding over on itself and not being welded up during later rolling or forging. It appears as a seam on the surface.

Martensitic. Steel with a microstructural constituent of long needle-like type, extremely hard, and formed when steel is quickly cooled from the hardening temperature.
Mechanical Hysteresis Loss. See Damping.

Notches. Sharp re-entrant angles in metals embodied in the design of a component or present as discontinuities in the surface.
Nuclei. The growing points at which cracks in a metal are initiated.

Parameter. A line or figure determining a point, line, figure or quantity in a class of such things, or a constant quantity in the equation of a curve.
Pearlite. A mechanical mixture or eutectoid consisting of the microconstituents cementite and ferrite in equal proportions. It gives a pearly appearance under the microscope.
Permanent Set. The permanent deflection of any structure after subjection to a full load: or in metallurgy, the extension left when the load is removed from a testpiece caused by the exceeding of the elastic limit.
Salt Bath. A bath composed of molten salts for heating steel in heat treatment. Salts of sodium, potassium, barium and calcium are mostly used.
Seams. Surface defects running lengthwise along a metal product, and appearing as grooves or striations of no great depth.
Segregations. Impurities in metal not evenly distributed.
Shot Peening. Throwing showers of steel or iron shot at

high velocity by compressed air against a metal surface to render it more fatigue resistant. It is not the same as shot blasting.

Sigma Phase. A hard brittle nonmagnetic compound in steels, often causing serious embrittlement at high temperatures.

Silky Fracture. A fracture having an extremely smooth fine dull grain, normally occurring in a metal of high ductility, such as low carbon steel.

Soft Skin. See Decarburization.

Space Lattice. The dimensional geometric pattern in which metallic atoms arrange themselves and upon which a crystal is built.

Stress Raiser. A condition causing considerable local increases of magnification of a stress. They are normally notches, grooves, or sharp changes in cross section, or inclusions, blowholes, folds, seams, etc.

Tough Fracture. Fracture of a ductile metal after heavy deformation.

Vitreous Fracture. Fracture in a fine grained nonductile metal.

Woody Fracture. A fibrous appearance of a fractured surface of a metal, caused by elongation of the separate crystals.

Workhardening. The increase in hardness and strength given to a metal component by deformation and mechanical working.

BIBLIOGRAPHY

Acta Met. Vol 7, pp 779-89, July 1963, 'Studies of Metal Crystals'
American Society for Metals, *Metals Handbook*, 'Heat Treating, Fatigue'
American Society for Testing and Materials, 1955-9, 1961-2, 1964, STP 9
American Society for Testing and Materials, Nov 1969, 'Fatigue at High Temperature'

Beddow, J. K. *Fatigue of Metals in Liquid Environments*, May 1968
Beelich, K. H. 'How Fatigue Affects Bolted Joints at High Temperature', *Metal Progress*, Aug 1970
British Standard Specification 3518, pt 1, 1962

Carter, C. J. and Celitti, R. A. 'How We Will Test Steels', *Metal Progress*, Oct 1969, p 215
Cina, B. 'Fatigue Strength of Cold Rolled 18-8 Cr Ni Steels, *Metallurgia*, May 1968
Clark, W. G. Jr. 'How Cracks Grow in Structural Steels', *Metal Progress*, May 1970, p 81

Encyclopedia of Workshop Practice, p 207
Forrest, P. G. *The Fatigue of Metals*, 1967
Forsyth, P. J. E. *The Physical Basis of Metal Fatigue*, 1969

Judge, A. W. 'Fatigue and Fatigue Cracking', *Engineering Materials*, Vol 2, pp 38-41, 120, 230, etc

Madayag, A. F. *Metal Fatigue, Theory and Design*, 1969, p 424

Myann. *Fatigue of Materials*, Melbourne, 1967
Metal Progress, June 1967, p 123, 'Drilling Platforms'
Sept 1969, Forum, pp 74-81, 'Fasteners in Aerospace Industry'
Metal Progress, July 1970, pp 100-2 'Nitriding Improves Fatigue Resistance'
Metal Progress, Feb 1971, p 95, 'Polarized Light Brings Out Details of Fracture Zones', C. S. Noritake, F. D. Walsh, E. C. Roberts
Little, R. E. *Machine Design*, June 1968, 'How to Prevent Fatigue Failure' p 154

National Lending Library for Science & Technology, 'Fatigue Strength of Metals'. Papers of Conference on Fatigue of Metals, USSR, 24-7 May, 1960

Plumbridge, W. S. and Ryder, D. A. *Metallurgical Review*, 3 Aug 1969

Ronson, L. and Knapton, A. G. 'Fatigue Strength of Metals' —Papers to 2nd Conference on Fatigue of Metals, Institute of Metallurgy, Sept 1968, p 68

Shives, T. R. and Bennett, J. R. 'Effect of Environment on Fatigue Properties of Selected Engineering Alloys', *Journal of Materials*, 3 no 3, Sept 1968, p 69 et seq.
Society of Automotive Engineers, *Fatigue Design Handbook*, New York, 1965
Strang, A. 'How Metals Break', *Metal Progress*, Feb 1970, p 42 et seq.

Welding Encyclopedia, Fatigue Limit, Stresses and Testing, p 223
Wiltzen, R. D. and Kives G. 'Impact Fatigue Testing of Titanium Alloys', *Journal of Metals*, Sept 1968, p 68

INDEX

Accelerated determination of fatigue limit, 111-12
Aircraft fasteners, 84-91
Alloy steels, 64-6
Aluminium:
 alloys, 77;
 die-castings, 76
Ambiguity, 98
Amsler testing machine, 54
Annealing, effect of, 58
Axial hydraulic closed loop machine, 98;
 thermal fatigue machine, 98-9

Basic fatigue terms, 116-20
Behaviour of components, 79-91
Bibliography, 124-5
Bolts, 84, 87-91
Brittle fractures, 21

Carbides, sintered, 67-8
Carbon:
 hot-finished steel, 60-1;
 steels, 60-1
Cartridge brass, 77
Cast components, 84
Causes of:
 corrosion, 45-7;
 corrosion fatigue, 43-4;
 fatigue failure, 37-41
Cleavage planes, 31-5
Cold-drawing steels, 63
Constructional steels, 61-2
Controlled interference-fit fastener, 84-5
Columbium alloy test, 102
Copper and copper alloys, 42-3, 76-7
Corrosion fatigue, 16-17, 41-7
Crack growth, 93-4

Cracking, 12, 14-16, 20-3, 93-4
Creep:
 effect of temperature on, 104-5;
 rupture ductility test, 103-4;
 test, high temperature, 103
Cross polarized light technique, 108-9
Crystal boundaries, 31-3
Crystalline fractures, 20
Cyclical loading, 14, 16
Cyclic stresses, 29-30

Damping capacity:
 27-8, 37, 113-14;
 and internal friction, 113-14
Decarburization, 39-41
Defects, 38-9
Directional flow, 51-2
Dowty testing machine, 72

Engine valves, 80-1
Environment, effect of, 49
Environment effects, 47-58
External medium and fatigue strength, 114-15

Fatigue:
 and hardened steels, 110-11;
 crack development, 110;
 mechanism of, 12-13;
 research, 96-7;
 resistance, metals for, 59-78
Fibre direction in steel parts, 111
Fibrous fractures, 20
Fractures, 12-13, 17, 19-21, 35-6
Fracture surfaces, 17
Fretting, 18
Friction, internal, 113-14

INDEX

Glossary, 116-24
Grain:
  growth, 34;
  size, 33

Hardness gradient, 51
Heat-resisting alloys, 74-5, 102
High purity nickel test, 102-3
High temperature:
  28-9;
  creep test, 103;
  low-cycle fatigue tests, 105
  tests, 101-7

Important tests, 98-9
Impurities, 32, 38-9
Induction hardening, 47
Internal stress, 22
Irons, 70-4

Liquid metals, effect of, 56-7
Loading conditions, 93-4
Lock nuts, 91
Low-cycle fatigue:
  life for pressure vessel, 105;
  test, 103;
  at high temperature, 105
Low temperature, effect of, 51-2
Lug failure, 18

Machine for studying corrosion
  fatigue, 115
Magnetic particle inspection, 39
Manganese steel, austenitic, 74
Matrix steels, 62
McQuaid-Ehn test, 34
Mean stresses, 23-4
Mechanism of:
  crack formation, 109-10;
  fatigue, 12, 13
Metalic and nonmetallic
  contents, effect of, 50-1
Metals for fatigue resistance,
  59-78
Metallurgical factors:
  24-41;

  terms, 120-4
Microstructure, 31-4, 41

Nickel, high purity test, 102-3
Niobium alloy test, 102
Nitempering, 49-50
Nitriding, 47
Nonferrous metals, 67
Nonmetallic inclusions, 19,
  39-41
Notch sensitivity of high
  strength steels, 113
Notches, 21, 24, 37-8

Oilholes, 21

Parameters, 85-6
Pins, 86-7
Pipes, 85
Pitting, 21
Plane bending, 16
Plated metals, 66
Powdered metals, 78
Pressure vessels, low-cycle
  fatigue life of steels for, —
Preventing fatigue failures, 23-4
Repeated bending:
  electromagnetic resonance
    machines, 98;
  test machines, 98
Rivets, 85
Rotating bending test machines,
  97

Scatter, 94-6
Schenk testing machine, 72
Shaft fracture, 53
Sintered carbides, 67-8
Size effect and corrosion fatigue,
  113
Sleeve bearings, 79-80
S-N curve, 55
Soft skin, effect of, 52, 56
Springs, 83-4
Stainless steels, 42, 66, 75-6,
  101-2

INDEX

Static shear failure, 19
Stress raisers, 21, 38
Striations, 13-14, 71
Structure energy theory, 110
Sulphide inclusions, 41

Testing:
  for fatigue, 92-107;
  machines, 54, 72, 97-101
Testpiece thickness, 57
Titanium:
  alloy fatigue properties, 106-7;
  and alloys, 68-70
Torsional:
  fatigue, 14;
  mechanical test machines, —
Treatment of fatigue cracks, 23
Tubes, 81-3

Ultrasonic tests, 99
USA research into fatigue failure, 109-15
USSR research into fatigue failure, 109-15

Vacuum heating, effect of, 57
Vibrational stresses, effect of, 52
Vibratory testing machine, new, 113-14

Welded joints, 81
Welding, effect of, 58
Woody fractures, 20
Workhardening, 12-13
Wrought steel, 64